Optimizing the Air Force Acquisition Strategy of Secure and Reliable Electronic Components

Proceedings of a Workshop

Committee on Optimizing the Air Force Acquisition Strategy of Secure and Reliable Electronic Components: A Workshop

Air Force Studies Board

Division on Engineering and Physical Sciences

The National Academies of
SCIENCES · ENGINEERING · MEDICINE

THE NATIONAL ACADEMIES PRESS
Washington, DC
www.nap.edu

THE NATIONAL ACADEMIES PRESS 500 Fifth Street, NW Washington, DC 20001

This activity was supported by Grant FA9550-14-1-0127 with the U.S. Air Force. Any opinions, findings, conclusions, or recommendations expressed in this publication do not necessarily reflect the view of any organization or agency that provided support for the project.

International Standard Book Number-13: 978-0-309-44518-4
International Standard Book Number-10: 0-309-44518-3
Digital Object Identifier: 10.17226/23561

Limited copies of this report are available from:

Air Force Studies Board
National Research Council
500 Fifth Street, NW
Washington, DC 20001
(202) 334-3111

Additional copies are available from:

The National Academies Press
500 Fifth Street, NW
Keck 360
Washington, DC 20001
(800) 624-6242 or (202) 334-3313
http://www.nap.edu

Printed in the United States of America

Suggested citation: National Academies of Sciences, Engineering, and Medicine. 2016. *Optimizing the Air Force Acquisition Strategy of Secure and Reliable Electronic Components: Proceedings of a Workshop*. Washington, DC: The National Academies Press. doi:10.17226/23561.

The National Academies of

SCIENCES · ENGINEERING · MEDICINE

The **National Academy of Sciences** was established in 1863 by an Act of Congress, signed by President Lincoln, as a private, nongovernmental institution to advise the nation on issues related to science and technology. Members are elected by their peers for outstanding contributions to research. Dr. Marcia McNutt is president.

The **National Academy of Engineering** was established in 1964 under the charter of the National Academy of Sciences to bring the practices of engineering to advising the nation. Members are elected by their peers for extraordinary contributions to engineering. Dr. C. D. Mote, Jr., is president.

The **National Academy of Medicine** (formerly the Institute of Medicine) was established in 1970 under the charter of the National Academy of Sciences to advise the nation on medical and health issues. Members are elected by their peers for distinguished contributions to medicine and health. Dr. Victor J. Dzau is president.

The three Academies work together as the **National Academies of Sciences, Engineering, and Medicine** to provide independent, objective analysis and advice to the nation and conduct other activities to solve complex problems and inform public policy decisions. The Academies also encourage education and research, recognize outstanding contributions to knowledge, and increase public understanding in matters of science, engineering, and medicine.

Learn more about the National Academies of Sciences, Engineering, and Medicine at **www.national-academies.org.**

COMMITTEE ON OPTIMIZING THE AIR FORCE ACQUISITION STRATEGY OF SECURE AND RELIABLE ELECTRONIC COMPONENTS: A WORKSHOP

ROBERT H. LATIFF, R. Latiff Associates, *Chair*
MICHAEL ETTENBERG, NAE,[1] Dolce Technologies
CRAIG L. KEAST, MIT Lincoln Laboratory
RANDAL W. LARSON, MITRE Corporation
TERRY P. LEWIS, Raytheon Company
CELIA MERZBACHER, Semiconductor Research Corporation
BERNARD S. MEYERSON, NAE, IBM
PAUL D. NIELSEN, NAE, Software Engineering Institute
STARNES E. WALKER, University of Delaware

Staff

JOAN FULLER, Director, Air Force Studies Board
CARTER W. FORD, Program Officer
MARGUERITE E. SCHNEIDER, Administrative Coordinator
STEVEN G. DARBES, Research Assistant
DIONNA C. ALI, Research Assistant

[1] NAE, National Academy of Engineering.

AIR FORCE STUDIES BOARD

[1] NAE, National Academy of Engineering.

Acknowledgment of Reviewers

This proceedings has been reviewed in draft form by individuals chosen for their diverse perspectives and technical expertise, in accordance with procedures approved by the Report Review Committee. The purpose of this independent review is to provide candid and critical comments that will assist the institution in making its published proceedings as sound as possible and to ensure that it meets institutional standards for objectivity, evidence, and responsiveness to the study charge. The review comments and draft manuscript remain confidential to protect the integrity of the deliberative process. We wish to thank the following individuals for their review of this proceedings:

Kent E. Devenport, Kansas City National Security Campus, Department of Energy,
Michael Fritze, Potomac Institute for Policy Studies,
Butler W. Lampson, NAS/NAE, Microsoft Research, and
Henry I. Smith, NAE, Massachusetts Institute of Technology.

Although the reviewers listed above have provided many constructive comments and suggestions, they were not asked to endorse the views presented at the workshop, nor did they see the final draft of the workshop proceedings before its release. The review of this workshop proceedings was overseen by Robert J. Hermann, NAE, Independent Consultant, Bloomfield, Connecticut, who was responsible for making certain that an independent examination of this workshop proceedings was carried out in accordance with institutional procedures and that all review comments were carefully considered. Responsibility for the final content of this proceedings rests entirely with the committee and the institution.

Contents

Acronyms

AFOSI	Air Force Office of Special Investigations
ASIC	application-specific integrated circuit
ASSP	application-specific standard product
BEOL	back end of line
CFIUS	Committee on Foreign Investment in the United States
CNCI	Comprehensive National Cybersecurity Initiative
COTS	commercial-off-the-shelf
CPU	central processing unit
DARPA	Defense Advanced Research Projects Agency
DIA	Defense Intelligence Agency
DMEA	Defense MicroElectronics Activity
DoD	Department of Defense
DoDI	DoD Instruction
DRAM	dynamic random-access memory
DSP	digital signal processor
FEOL	front end of line
FPGA	field-programmable gate array
FY	fiscal year
GPU	graphic processing unit
HW	hardware
IA	information assurance
IARPA	Intelligence Advanced Research Projects Activity
IC	integrated circuit; Intelligence Community
ICT	Information and Communications Technology
IP	intellectual property
JFAC	Joint Federated Assurance Center
KCNSC	Kansas City National Security Campus
MPU	multi-core processing unit
MTO	Microsystems Technology Office
NDAA	National Defense Authorization Act

NDIA	National Defense Industry Association
NEA	Nuclear Enterprise Assurance
NIST	National Institute of Standards and Technology
NNSA	National Nuclear Safety Administration
NSA	National Security Agency
NSE	Nuclear Security Enterprise
NSS	National Security Space
NSWC	Naval Surface Warfare Center
OEM	original equipment manufacturer
OSD	Office of the Secretary of Defense
PPP	program protection planning
R&D	research and development
RFP	request for proposal
ROI	return on investment
RSA	Rivest-Shamir-Adleman
SCRM	Supply Chain Risk Management
SES	Senior Executive Service
SHIELD	Supply Chain Hardware Integrity for Electronics Defense
SMC	Space and Missile Systems Center
SOC	system on a chip
SP	Special Publication
SRAM	static random-access memory
SSE	Security Systems Engineer
SWaP	size, weight, power
TAC	Threat Assessment Center
TAPO	Trusted Access Program Office
TIC	Trusted Integrated Chips
WTA	Weapon Trust Assurance

Overview

In 2012, the National Defense Authorization Act (NDAA), section 818, outlined new requirements for industry to serve as the lead in averting counterfeits in the defense supply chain.[1] Subsequently, the House Armed Services Committee, in its report on the Fiscal Year 2016 NDAA, noted that the pending sale of IBM's microprocessor fabrication facilities to Global Foundries created uncertainty about future access of the United States to trusted state-of-the-art microelectronic components and directed the Comptroller General to assess the Department of Defense's (DoD's) actions and measures to address this threat.[2,3,4] In this context, the Deputy Assistant Secretary of the Air Force (Science, Technology, and Engineering) requested that the Air Force Studies Board of the National Research Council[5] convene a workshop to facilitate an open dialogue with leading industry, academic, and government experts to (1) define the current technological and policy challenges with maintaining a reliable and secure source of microelectronic components; (2) review the current state of acquisition processes within the Air Force for acquiring reliable and secure microelectronic components; and (3) explore options for possible business models within the national security complex that would be relevant for the Air Force acquisition community. This report summarizes the results of a workshop held on March 16-18, 2016, in Washington, D.C., which brought together experts from government, industry, and academia to address these issues.

THE MICROELECTRONICS LANDSCAPE

During the "dawn" of the semiconductor industry in the 1970s, the focus was on ensuring that specific, required functionality was available through the design, fabrication, and production of application-specific integrated circuits (ASICs) and mass produced computer memories. Since then, advances in device speed, increased processing power and throughput, lower electrical power consumption, vast increases in device volume production, and ingenious, complex designs have enabled numerous new applications and enormous improvements. This rate of technological advance is expected to continue and perhaps accelerate as new substrate materials are introduced.[6]

Because electronic components in many national security systems are designed and intended to last for long periods in sometimes in harsh environments, testing to assure that the parts will indeed function properly and reliably, under all conceivable operational conditions, and function only as designed, becomes

[1] National Defense Authorization Act for Fiscal Year 2012 (P.L. 112-81).

[2] J. Lipsky, "IBM-GlobalFoundries Deal Finalized," *EETimes.com*, July 1, 2015, http://www.eetimes.com/document.asp?doc_id=1327029.

[3] National Defense Authorization Act for Fiscal Year 2016, H.R.1735, 114th Congress, https://www.congress.gov/bill/114th-congress/house-bill/1735, accessed April 17, 2016.

[4] Global Foundries is an international company headquartered in Santa Clara, California. It is owned by the Mubadala Development Company, a wholly-owned investment vehicle of the Government of Abu Dhabi in the United Arab Emirates.

[5] Effective July 1, 2015, the institution is called the National Academies of Sciences, Engineering, and Medicine. References in this report to the National Research Council (NRC) are used in a historical context to refer to activities before July 1.

[6] A recent Aerospace Corporation study (TOR-2015-00473) included a summary of "Technology Challenges by 2025."

challenging.[7] The design of such tests requires intimate knowledge of the device operation and requires sophisticated testing techniques and equipment. Government program managers, program executive officers, and agency leaders are faced with the choice of either using commercial-off-the-shelf (COTS) devices—which may or may not support their requirements—and accept unknown risks; or they will have to make significant investments in test and certification technologies to validate operating parameters.

Complicating this situation further is the steadily eroding U.S. involvement in the design and manufacture of necessary electronic devices, and a concomitant decrease in domestic expertise and understanding of reliability and the risks to systems associated with such complexity. As a result of this erosion, there may not be a domestic microelectronics workforce capable of generating the required security and reliability information the government would require to appropriately analyze and advise program managers about the attendant system risks of microelectronic components.[8]

ORGANIZATION OF THE WORKSHOP

Workshop briefings included information on (1) DoD's strategy for acquiring secure and reliable microelectronic components, (2) the needs of the nuclear weapons enterprise, (3) Air Force processes to gather reliable and secure information, (4) Defense MicroElectronics Activity's (DMEA's) new role as the sole manager of the Trusted Access Program Office (TAPO), (5) Defense Advanced Research Projects Agency and Intelligence Advanced Research Projects Activity technology research and development programs to insure that obtained parts are secure, and (6) the important role of standards in the manufacture and testing of secure and reliable microelectronic components. Importantly, briefings by industry shed light on the economics of electronics manufacturing and highlighted the pros and cons of government ownership of trusted foundries.

One of the issues that was raised repeatedly during presentations was the prohibitive cost associated with dedicated state-of-the-art foundries producing secure and reliable microelectronic components for national security systems. A few participants noted that a main reason associated with the high cost of producing these items is the relatively low volume of items required by DoD and the Intelligence Community in comparison with the commercial marketplace. More than one speaker from government and industry noted that without a reasonable market, industry will find it difficult to support a program based entirely on producing low-volume trusted components for government systems. Other participants commented that another barrier for industry support of producing low-volume trusted components for the government is the burdensome accreditation process the government uses to determine whether a potential supplier is trustworthy. For example, DMEA performs an accreditation process via a Cooperative Research and Development Agreement that allows DMEA to work with a potential supplier every 2 years. More than one participant asked the speaker from DMEA why a supplier would not want to be accredited. Several reasons were provided, including cost, return on investment, fear of not passing the screening, and the potential market share not matching a company's business model. Yet another participant commented that, on an anecdotal level, existing trusted suppliers who were not receiving requests for trusted fabrication prior to the IBM/Global Foundries sale are now seeing an increase in inquiries as a result of the sale. Finally, concerns about the burdens on industry associated with the International Traffic in Arms Regulations, as well as the complex U.S. government acquisition and contracting process, were mentioned by more than one participant during the course of the workshop.

[7] There are distinct approaches involved when it comes to testing components for security as opposed to testing them for reliability when a suspicious malicious actor is not involved.

[8] U.S. Government Accountability Office, *Trusted Defense Microelectronics: Future Access and Capabilities are Uncertain*, GAO-16-185T, Washington, D.C., October 28, 2015.

REPORT ORGANIZATION

Chapter 1 provides a broad contextual background that includes challenges related to current government policies and technological advancements in the area of secure and reliable electronic components in government national security systems. Even though attribution to individual speakers or workshop participants is not provided, this section of the report should not be seen as consensus views of the wide representation of views presented throughout the workshop. Chapter 2 goes on to describe the dialogue that occurred at the workshop, followed by Chapter 3, which provides abstracts of speaker presentations. Appendixes are provided at the end of the report and include the following items: (1) workshop terms of reference, (2) brief biographies of the workshop committee members, (3) speakers and attendees list, (4) suggested terms of reference for a follow-on study, and (5) a summary presented by Bernard Meyerson of his thoughts on the projected advancements of existing technology. *This proceedings summarizes the views expressed by individual workshop participants. While the committee is responsible for the overall quality and accuracy of the proceedings as a record of what transpired at the workshop, the views contained in the proceedings are not necessarily those of all workshop participants, the committee, or the National Academies of Sciences, Engineering, and Medicine.*

1

Workshop Context and Issues

To help provide both context and focus for the workshop, several members of the workshop organizing committee (i.e., Craig Keast, Michael Ettenberg, Robert Latiff, Bernard Meyerson, and Paul Nielsen) framed the workshop discussions by highlighting that advanced electronic devices are critical for all U.S. national security systems, military or intelligence related. The increasing demands for performance of these systems have led to the adoption of ever more sophisticated devices for sensing, computing, control, and other critical functions. For several decades, the technologies for making integrated circuits and microprocessors followed Moore's Law. This "scaling" had the virtuous benefit of making products that were faster, better (i.e., more functional and power efficient), and cheaper, stimulating an enormous information technology industry. Although the cost per transistor steadily decreased, the cost to build foundries for such devices grew in a commensurate fashion; a state-of-the-art foundry costs on the order of $5 billion to build.[1] Much of the manufacturing of this nature is in Asia. U.S. aircraft, missiles, ships, and ground vehicles, as well as radars and other sensors, depend on access to electronics components that are known to be reliable and to perform as designed. The primary goal of program managers and engineers in national security programs is to assure mission success of weapon systems, and access to reliable and trusted microelectronics are essential to assuring that success.

Many of the technologies critical to national security are dependent on leading-edge semiconductors and microelectronic devices that, in many cases, do not have a commercial market (see Figure 1-1). Another school of thought, expressed by one workshop participant, is that leading-edge semiconductors can only be made in high-volume commercial fabrication facilities.

As described by several of the workshop participants (e.g., Kristen Baldwin, Jimmy Goodrich, Terry Lewis, Bernard Meyerson, Celia Paulson, and Dustin Todd) over the 3-day workshop, the acquisition of electronic devices is a complex process that often defies simplification. It includes everything from the sourcing of raw materials, to wafer manufacture, to component design, to software development, to assembly, to testing and certification. The continued and accelerating globalization of the microelectronics industry presents national security program designers with a challenge of how to ensure that electronic components operate as designed. Off-shoring of parts manufacture, decreased Department of Defense (DoD) influence on the industry due to a small comparative demand, and diminished U.S. expertise are all contributing to a growing inability to either understand or assure system security and reliability. The electronics supply chain is complex and has many points within it that can present problems for the ultimate security and reliability of its products. Increasingly, end users demand to know the "pedigree" of the parts they are acquiring for high-priority national security systems. In general, it may be possible to insure greater supply chain trust and reliability of parts by implementing stronger community policies, information sharing on issues and solutions, and coordinated investments in research and development (R&D).

As shown in Figure 1-2, DoD identifies a spectrum of risks to the electronics supply chain. They include (1) quality escapes due to inadequate design or manufacturing quality control; (2) reliability failures; (3) insertion of fraudulent or counterfeit products; (4) insertion of malicious hardware, software, or computer

[1] Christopher Mims, "The High Cost of Upholding Moore's Law," *MIT Technology Review*, April 20, 2010.

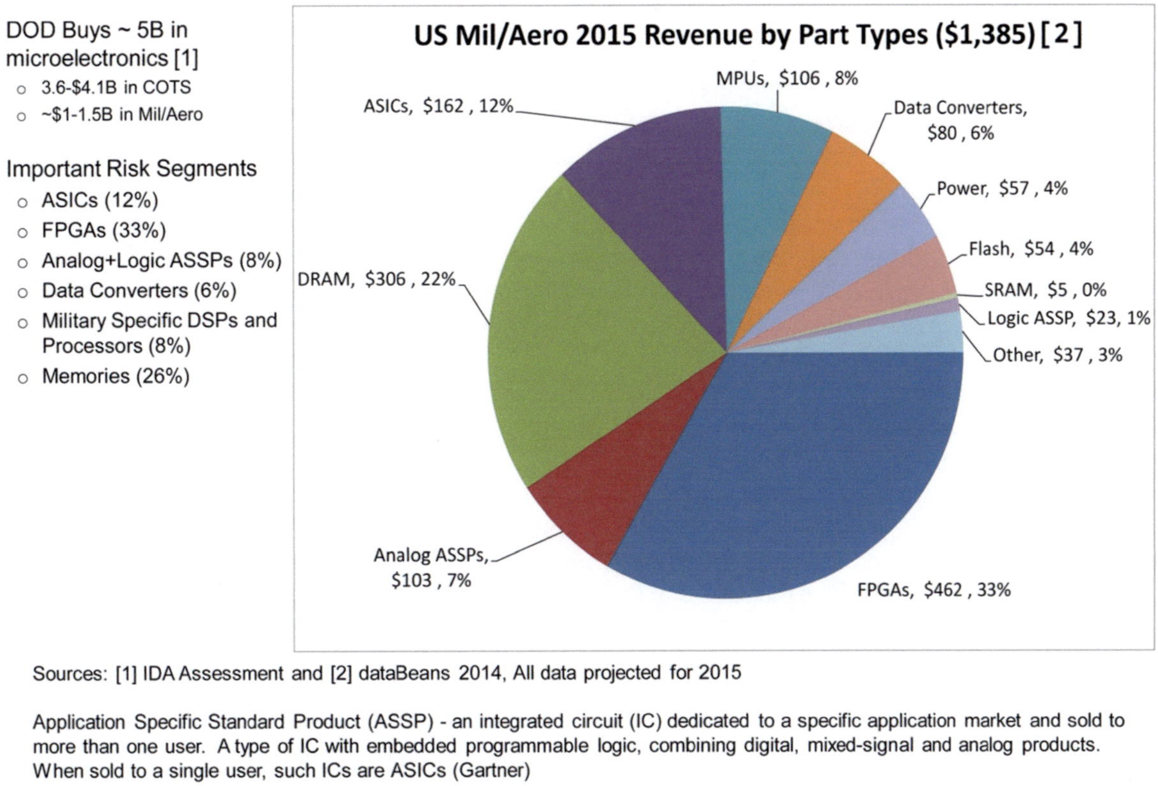

FIGURE 1-1 Microelectronics in Department of Defense systems. NOTE: Acronyms are defined in the front matter. SOURCE: Brian Cohen, Institute for Defense Analyses, presentation to the workshop on March 18, 2016.

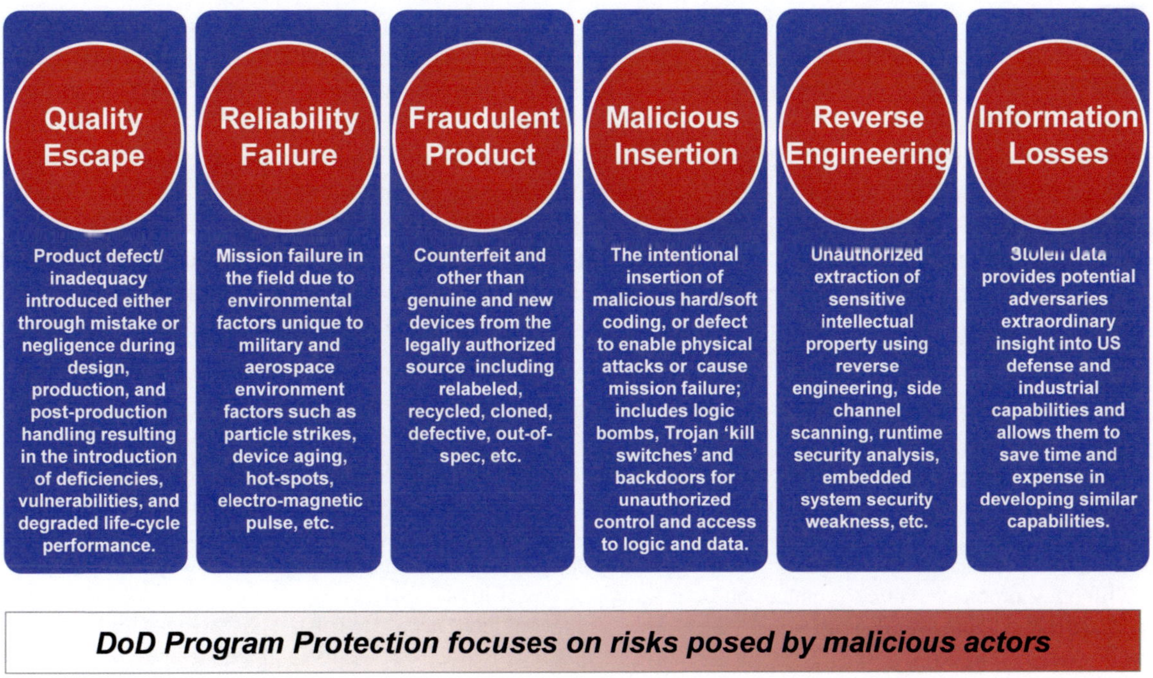

FIGURE 1-2 Spectrum of supply chain risks. SOURCE: Kristen Baldwin, Acting Deputy Assistant Secretary of Defense for Systems Engineering and Principal Deputy Assistant Secretary of Defense for Systems Engineering, presentation to the workshop on March 16, 2016. Distribution Statement A—Approved for public release by DOPSR; SR#15S-1541 applies. Distribution is unlimited.

code intended to cause mission failure; (5) reverse engineering of sensitive intellectual property or government information; and (6) outright theft of information that allows adversaries to achieve capabilities they would not otherwise obtain.

DoD's strategy to ensure that critical and sensitive electronics remain viable includes (1) protection of microelectronics designs and intellectual property; (2) advanced hardware analysis capabilities; (3) physical, functional, and design verification and validation; and (4) a new trust model that leverages commercial state-of-the-art capabilities. As an example of this layered approach, the federal government has initiated investments in the development of new, trusted photomask capabilities, tools to enhance the ability to detect flaws, and increased academic and industry research in this area.[2,3] One workshop participant noted that the Trusted Access Program Office (TAPO) also plays a very important function in DoD strategy. TAPO currently manages the trusted part contract with Global Foundries U.S. and is speaking with other fabrication facilities and companies that are manufacturing field-programmable gate arrays (FPGAs) to develop trusted access solutions.

As described by at least one participant during the workshop, prior to the past two decades, the U.S. government had generally enjoyed a mutually beneficial relationship with its supply chain where the government could be assured of acquiring high reliability and state-of-the-art technologies, and suppliers could be assured of benefitting from the results of their R&D investments within a future commercial market. Today, trusted domestic suppliers increasingly find it necessary to forge and accept commitments with what the government may consider non-trusted sources to ensure their own corporate survival within a highly competitive global marketplace. A few participants commented that there are many reasons for this U.S.-supplier marketplace transition. Among them, and perhaps most relevant to the part of the "trusted" microelectronics industry dedicated to the government user, is the near-total loss of on-shore domestic capabilities to fabricate complex, state-of-the-art, highly reliable electronic parts.[4] Another participant commented on the equally important concern stemming from an increasing dependency by the government on the obsolete electronic parts "grey market" where a counterfeit sub-industry has firmly established itself.

Throughout the workshop, several speakers and attendees reinforced the belief within the defense community that the trusted supplier or supply chain is the foundation of assurance for microelectronic parts. Without it, alternative methods to understand the integrity of the product need to be applied and may not achieve the same level of confidence as that won with the trusted supplier/supply chain. However, several participants noted that in lieu of having a trusted supplier or an end-to-end trusted production flow for certain microelectronics, there are efforts underway today to create what are thought to be acceptable alternatives, including broadening the acceptable use of otherwise untrusted sources. Some refer to this concept as establishing "tiers" of trust.[5] Another method to reduce costs for obtaining assurance in lieu of a trusted supply chain that encompasses all electronic components is one that instead focuses the trust requirements only on mission critical parts. Unfortunately, as noted by several participants, more traditional approaches to assuring trust may prevail during more robust financial environments; however, today's budget realities and

---

[2] "A photomask is a tool used for production of components including electronic devices (semiconductors), displays, PCB, and MEMS. It is a master copy for the patterning. Photolithography is used to form PCB circuits and display patterns. Photomasks are used to transfer the patterns on the baseplates. A photomask acts just like "negative film" in photography, and that makes the baseplates "printing paper" (See Filcon Photomask, "What is a Photomask?" http://filcon-photomask.com/en/product/photomask.php, accessed July 7, 2016).

[3] "In the event that the GF Trusted Foundry closes, DoD would lose access to trusted photomasks for leading-edge designs" (Kristen Baldwin, Acting Deputy Assistant Secretary of Defense for Systems Engineering and Principal Deputy Assistant Secretary of Defense for Systems Engineering, presentation to the workshop on March 16, 2016).

[4] There is U.S.-based, leading-edge manufacturing capability (e.g., Intel). The lack of a leading-edge technology supplier in the United States is more complicated than "they are all off-shore." The current business model requires extremely large volumes, and this does not align with current government procurement practices and programs.

[5] The Potomac Institute for Policy Studies is currently undertaking a major 1-year study for DoD to develop such a "tiered" system of trust.

limited trusted supplier base for certain devices are forcing managers to take greater programmatic risks.[6] The risks incurred from the acquisition of bad electronic parts from a non-trusted source vary across the spectrum of technical failure modes. Risk impacts that may be realized can be mission-ending, disrupting failures, or life-compromising reliability issues. A poorly managed supply chain offers several points of intrusion or entry for bad actors to insert malicious or counterfeit hardware, software, or firmware.[7] As government systems age, their growing dependency on obsolete parts subjects the buyer to a large, global vendor market of non-OEMs (original equipment manufacturers). An example may be that a vendor is based in the United States with claims of having a desired part, yet may, in fact, reach-back for the parts to unknown sources. Other bad actors may have interests in disrupting a system or compromising its mission life and may have very sophisticated techniques to fool the unsuspecting intake engineer into accepting the product.

An example of an organization that pays attention to electronics obsolescence and to supplier trust accreditation is the Defense MicroElectronics Activity (DMEA). In his presentation and the ensuing discussion, Dan Marrujo from the DMEA described the role of his organization in addressing many of the challenges that were highlighted during the discussions with Kristen Baldwin. DMEA is a key element in the assurance of continued access to obsolete parts and in certifying suppliers for trusted status. One element of the DMEA mission is to re-engineer and manufacture advanced microelectronics parts no longer available to program managers through their industry partners or through other standard commercial sources. Also, DMEA is currently the program manager for the DoD Trusted Foundry Program. Among other tasks, the program negotiates and manages trusted access contracts with state-of-the-art fabrication facilities (e.g., GlobalFoundries U.S.) and accredits sum-of-the-parts microelectronics companies for trust. DMEA accredits suppliers' processes in the areas of integrated circuit design, aggregation, broker, mask manufacturing, foundry, post processing, and packaging/assembly and test services. DMEA is a member of the Joint Federated Assurance Center (JFAC) Working Group. Other members include the Office of the Under Secretary of Defense (Acquisition, Technology, and Logistics), the DoD Chief Information Officer, Military Departments, the Missile Defense Agency, the National Security Agency, the National Reconnaissance Office, and the Defense Information Systems Agency. The JFAC, created by the Deputy Secretary of Defense, identifies, promotes, and facilitates access to hardware and software assurance (i.e., verification and validation) capabilities across the DoD and other federal agencies throughout the system life cycle.)

Several of the workshop participants commented that the ongoing challenge in microelectronics evolution is sheer complexity: logic devices such as application-specific integrated circuits (ASICs) or FPGAs are so complex that determining how to best verify the integrity of the product when the parts may be fabricated in an untrusted foundry has been a more recent, and increasing, concern for programs. ASICs and FPGAs often provide the logic required to drive a critical function. They need to be reliable and tamper-free. Having a trusted supplier and ensuring end-to-end trusted production flow are not achievable goals for some programs While many studies and innovative technical approaches are under way today to determine methods for achieving some level of confidence that parts will be reliable and can be trusted, no definitive comprehensive approach has been identified to date.

While understanding and attempting to assure the integrity of the supply chain is critical, at the end of the day, designers and system developers need to convince themselves that the delivered electronic products will actually function as advertised, for the length of time needed by the mission, under the conditions expected, and be free from tampering or malicious content. To do so requires rigorous testing and a well-designed certification scheme. Maintaining and assuring the complete integrity of the supply chain is difficult because of the complexity and interconnectedness of the supply chain elements. Items include the raw materials, development tools, facilities and their integrity (production and storage), and the complex machines used to produce parts and their associated programming.

---

[6] One workshop participant noted that there are 72 suppliers on the DMEA accreditation list. This is not a small number, but only a limited number are, in fact, being used for U.S. government needs.

[7] See U. Guin, D. DiMase, and M. Tehranipoor, A comprehensive framework for counterfeit defect coverage analysis and detection assessment, *Journal of Electronic Testing* 30(1):25-40.

The contractor community drives, and is driven by, system performance requirements. They need to balance demands for increased performance (e.g., decreased feature size, increased density) with strict security and reliability guidelines. Prime defense contractors have serious concerns about the health of the available industrial base, as well as the ability to obtain quality parts. Significant resources are expended by the industry in quality assurance, as most electronic component suppliers are now off-shore. The supply chain, and the ability to assure its integrity, becomes a very important issue for weapon system developers and electronic component manufacturers. Industry watchers are concerned with an accelerating rate of consolidation and closures that are taking place within the manufacturing sector.

In summary, the workshop presentations and discussions highlighted the observation that the national security electronics industrial base is being pulled in different directions. On the one hand, they are at the mercy of the electronics manufacturers and suppliers. On the other hand, the government program offices are making performance demands, security demands, and reliability demands that the industrial base is increasingly unable to guarantee. The problem is exacerbated by diminishing support by the government for expensive and unique test facilities and inconsistent requirements from the system designers. The industry is looking to the government for leadership and guidance and, in its absence, is having to make tough, sometimes non-optimum, choices. The industrial base for national security systems has significant concerns with the state of the microelectronics industry and its ability to supply the kind of high-quality, high-reliability systems needed for their products.

2

Workshop Discussions and Key Themes

Over the course of the 3-day workshop, there were numerous topics raised by speakers and brought up during related discussions. These topics and discussions are organized, roughly, according to the workshop terms of reference (provided in Appendix A). Finally, there are contained in each of the following sections certain key themes that arose during the workshop across multiple presentations and associated discussions, and these are highlighted below.

CURRENT TECHNOLOGICAL AND GOVERNMENT POLICY CHALLENGES

Current Department of Defense (DoD) policy guidance pertaining to secure and reliable microelectronic components is covered by DoD Instruction (DoDI) 5200.44 and DoDI 4140.67.

> Establishes policy and assigns responsibilities to minimize the risk that DoD's warfighting mission capability will be impaired due to vulnerabilities in system design or sabotage or subversion of a system's mission critical functions or critical components by foreign intelligence, terrorists, or other hostile elements.[1]

> Establishes policy and assigns responsibilities necessary to prevent the introduction of counterfeit materiel at any level of the DoD supply chain.[2]

The Defense MicroElectronics Activity (DMEA) is the sole manager of the Trusted Access Program Office (TAPO) that is responsible for ensuring that trusted microelectronics are available for U.S. national security systems.[3] The speaker from DMEA noted that his organization is primarily interested in DoDI 5200.44—specifically, the requirement to use trusted foundries and suppliers for application-specific integrated circuits (ASICs). A participant from the Office of the Secretary of Defense noted that DoDI 5200.44 requires and promulgates acquisition programs to use only ASICs that have been designed, fabricated, and packaged by suppliers that have been "trust" accredited by DMEA. Importantly, however, one participant noted that the current government policies only cover ASICs and do not address other commercial-off-the-shelf (COTS) electronic components, which make up the majority of microelectronics used in DoD mission-critical systems. DoDI 5200.44 also authorizes the Air Force Office of Special Investigations (AFOSI) to investigate and provide threat reports upon request. These threat reports are created and disseminated through the Defense Intelligence Agency's (DIA's) Threat Assessment Center (TAC).[4]

---

[1] Department of Defense, "Protection of Mission Critical Functions to Achieve Trusted Systems and Networks (TSN)," DoDI 5200.44, November 5, 2012. http://www.dtic.mil/whs/directives/corres/pdf/520044p.pdf.

[2] Department of Defense, "DoD Counterfeit Prevention Policy," DoDI 4140.67, April 26, 2013, http://www.dtic.mil/whs/directives/corres/pdf/414067p.pdf.

[3] The TAPO was established in 2006 based on the recommendations of Defense Science Board, *Task Force on High-Performance Microchip Supply*, 2005, http://www.acq.osd.mil/dsb/reports/ADA435563.pdf.

[4] According to the speaker from AFOSI, the Intelligence Community does not have a policy directive equivalent to DoDI 5200.44.

USD(I)	DoDI 5200.39 - Critical Program Information (CPI) Identification and Protection Within Research, Development, Test, and Evaluation (RDT&E)
L&MR	DoDD 5134.12 Logistics & Materiel Readiness Organization
DSS	National Industrial Security Program (NISP) - Executive Order 12829 Foreign Ownership, Control or Influence (FOCI)
CIO	Protection of Mission Critical Functions to Achieve Trusted Systems and Networks (TSN)DoDI 5200.44
MIBP	Ensure robust, secure, resilient, and innovative industrial capabilities upon which the DoD can rely to fulfill the Warfighter's requirement
DoDD 5000	DoDI 5000.02 (7Jan 2015) USD(AT&L): Operation of the Defense Acquisition System
DoDI 4140.01	Establishes policy and assigns responsibilities for management of materiel across the DoD supply chain.
Acquisition	DoDI 5000.02, Jan 7, 2015, Operation of the Defense Acquisition System All of the Above; Affordability; Sequestration, Ind Base, Budgets
DLA	QPL/QML

FIGURE 2-1 Current government policies pertaining to secure and reliable microelectronic components. SOURCE. Dave Davis, Chief Engineer, Air Force Space and Missile Systems Center, presentation to the workshop on March 17, 2016.

The chief engineer from the Air Force Space and Missile System Center (SMC) agreed and affirmed that SMC follows DoDI 5200.44 in the areas of getting DIA TAC reports for risk assessments, counterfeit prevention, and the use of DMEA-accredited ASICs. A DMEA representative noted that there is widespread knowledge of the DoDI 5200.44 policy, but not necessarily the *definition* of the policy. A speaker from the National Defense Industry Association (NDIA) confirmed this view by stating that there is a knowledge gap in government of the requirements of 5200.44 and that there is a need to educate the acquisition community on 5200.44.[5] Finally, Figure 2-1 provides examples of the organizations and policies addressing the multiple missions and solutions required to address integrity assurance in microelectronic components used in DoD national security and weapons systems, according to multiple participants.

Finally, a key limitation with respect to government policies in the area of microelectronics is the time involved in drafting and implementing new DoD-wide policies—specifically, it can take up to 2 years to write and 2 years to implement new policies, according to a participant from the Office of the Secretary of Defense (OSD). At the same time, this participant agreed that more guidance in this area will help.

In addition to DoDI 4140.67 and DoDI 5200.44, DoDI 5000.02 requires government and industry program managers to employ system security engineering and prepare and maintain a program protection plan (PPP) throughout the acquisition life cycle of a weapon system.[6] According to the speaker from OSD, a PPP requires the identification of critical components in a weapon system and associated risk assessment based on threats, vulnerabilities, and mission criticality. According to the speaker from SMC, they have been

[5] The following section provides examples for better government and industry collaboration.

[6] Department of Defense, "Operation of the Defense Acquisition System," DoDI 5000.02, January 7, 2015, http://www.dtic.mil/whs/directives/corres/pdf/500002p.pdf.

performing rigorous program protection and countermeasures of the supply chain out of necessity for decades, due to the demands of space systems. Lessons learned from SMC's approach to monitoring suppliers may be applicable to others, including the Joint Federated Assurance Center (JFAC). In response to recent releases where the DoD, NDIA, and the National Institute of Standards and Technology now specify that systems engineers are to provide program protection planning, the speaker noted that SMC's systems engineers have historically performed this function. The speaker noted that requirements call for a security systems engineer (SSE) who performs the oversight of the program protection effort and is aligned with most of the activities currently being performed by SMC systems engineers.

Lastly, embedded systems were noted by multiple participants to be the next big policy issue in the area of secure and reliable microelectronics. The issue is that third-party providers who supply the embedded systems are not scrutinized by the DoD program protection policies being imposed on the discrete component providers. An example that was discussed during the presentation from AFOSI related that the provenance and design documentation, which is considered intellectual property (IP) by the owners of the embedded systems, is rarely provided. This results in components from suppliers that are unspecified to DoD being placed in systems to perform the most critical functions—for example, random number generators. In fact, the majority of microprocessor design products may be from third-party providers.[7]

Key Theme 1—DoDI 5200.44

As noted by multiple speakers and participants (e.g., Kristen Baldwin, Daniel Marrujo, and Michael Lyden), DoDI 5200.44 has had a big impact on DoD's approach to Supply Chain Risk Management (SCRM), including (1) enforcing an updated approach to program protection planning; (2) expanding the mission of DMEA; (3) requiring ASICs to be supplied by a trusted foundry; (4) enabling AFOSI to investigate domestic companies and U.S. persons for supply chain threats; (5) requiring testing to evaluate the trustworthiness of hardware and software components; and (6) requiring more rigor in the prevention and detection of counterfeits.

Key Theme 2—Program Protection Policies

Several presentations (e.g., David Davis, Kent Devenport, Holly Dunlap, John Boyens, and Celia Paulson) revealed that program protection imposed by "top down" policy requires "bottom up" implementation in order for the intent of integrating trust, through verifiable confidence in the integrity of the hardware, firmware, and software components, to be realized. The acquisition reality is that if a fool-proof trusted component was provided, who would be required to use it and by what evidence could it be accepted if not documented by these policies and processes?

[7] Microprocessors are a security concern because of the impact they have on system operations, and the design complexity involved, which makes detection of hidden or unwanted functions to be extremely difficult. But there are many other complex devices vulnerable as well, such as FPGAs, ASICs, memory, and random number generators, which play a crucial role in intense computations, such as cryptographic functions.

Current Technology Capabilities to Detect Fraud and Counterfeits

Counterfeit and clone components are increasingly an issue that the DoD is facing with respect to secure and reliable microelectronics and were the main topics addressed by speakers from AFOSI and the Naval Surface Warfare Center (NSWC). According to presentations, the United States is losing critical IP due to globalization. Key issues the government is facing include (1) clones fabricated in unknown foundries that mimic the operation of authentic parts and (2) replications derived from stolen IP that instead are reverse-engineered *with potentially altered function*. A participant noted that the examples shown in the presentations were of older technologies and asked if counterfeiting is more of an issue with older technologies. The speaker from NSWC replied that counterfeiters are rapidly keeping up with advances in technology. Relatedly, the speaker from AFOSI noted that the Air Force is the largest consumer of old and obsolete technologies and that there are no parts that are beyond interest of counterfeiters. Upwards of 50 percent of Air Force sustainment parts originate in the grey market.

Pertaining to the issues raised in the NSWC presentation, there is inherent risk in looking for counterfeits due to false positive test results that have been observed in some test methods. Some participants noted that variations in chip measurements are criteria for binning of chips per performance measured (an accepted practice). However, these participants noted that measuring these variations is not a criteria for detection of counterfeits, making detection of real counterfeits difficult. The speaker from NSWC cited other methods that are more reliable indicators of counterfeits, such as principal component analysis and vector impedance measurements (see Figure 2-2).[8] The participants from the Department of Energy's Kansas City National Security Campus (KCNSC) stated that Sandia National Laboratories performs all of the testing for Nuclear Enterprise Assurance. One participant stated that establishing the trustworthiness of field-programmable gate arrays (FPGAs) presents multiple concerns, including the following: (1) threats due to malicious insertion, (2) vulnerabilities in programming, (3) complexity in detection methods, and (4) prominence of counterfeits. Another participant commented that a lot of the verification and evaluation tools used for space systems are classified and asked, How can we share these with the broader community and industry? A participant replied that DoD is working on a classification guide for the JFAC for how to share information on vulnerabilities.

Counterfeit parts are easier to make and sell because they do not necessarily have to work in the system under all conditions, as did the original part. They could also contain circuitry that has malicious content that can be activated at some point in the future. The speaker from NSWC noted that, while there are a large number of physical investigative techniques, ranging from simple visual inspection through destructive analysis using scanning electron microscopy, this is a slow and expensive process because it requires having knowledge of the intended design, the use of "golden units" for comparison, and extensive training. Having knowledge of the origin of parts is preferred because it provides legitimacy to the claim of authenticity. Program managers can avoid purchasing parts from after-market suppliers and distributors, however, with system lifetime buys of mission-critical parts at the outset of a program, which enables procurement from the original component manufacturers (OEMs) during production of those parts.

Many workshop participants were encouraged by the innovative and promising initiatives that the Defense Advanced Research Projects Agency (DARPA) either had under way or was starting to ensure the provenance of future integrated circuit parts. For example, the Supply Chain Hardware Integrity for Electronics Defense (SHIELD) program (a taggant) is intended to impose a cost and time asymmetry on the adversary.[9] The taggant is embedded in the package material of the integrated circuit. These will work—as long as they are affixed to legitimate hardware—and are cost effective, but they do not solve the software side of the problem. Software integrity is a more immediate, and probably larger, problem

[8] *Impedance* is the effective resistance of an electric circuit or component to alternating current, arising from the combined effects of ohmic resistance and reactance (*Oxford Dictionaries*, http://www.oxforddictionaries.com/us/definition/american_english/impedance, accessed July 7, 2016).

[9] A taggant is a unique signature found in an electronic component similar to strips found in currency notes to deter counterfeiters. For additional information on different forms of taggants, see Microtrace, "What is Taggant?," http://www.microtracesolutions.com/taggant-technologies/, accessed June 27, 2016.

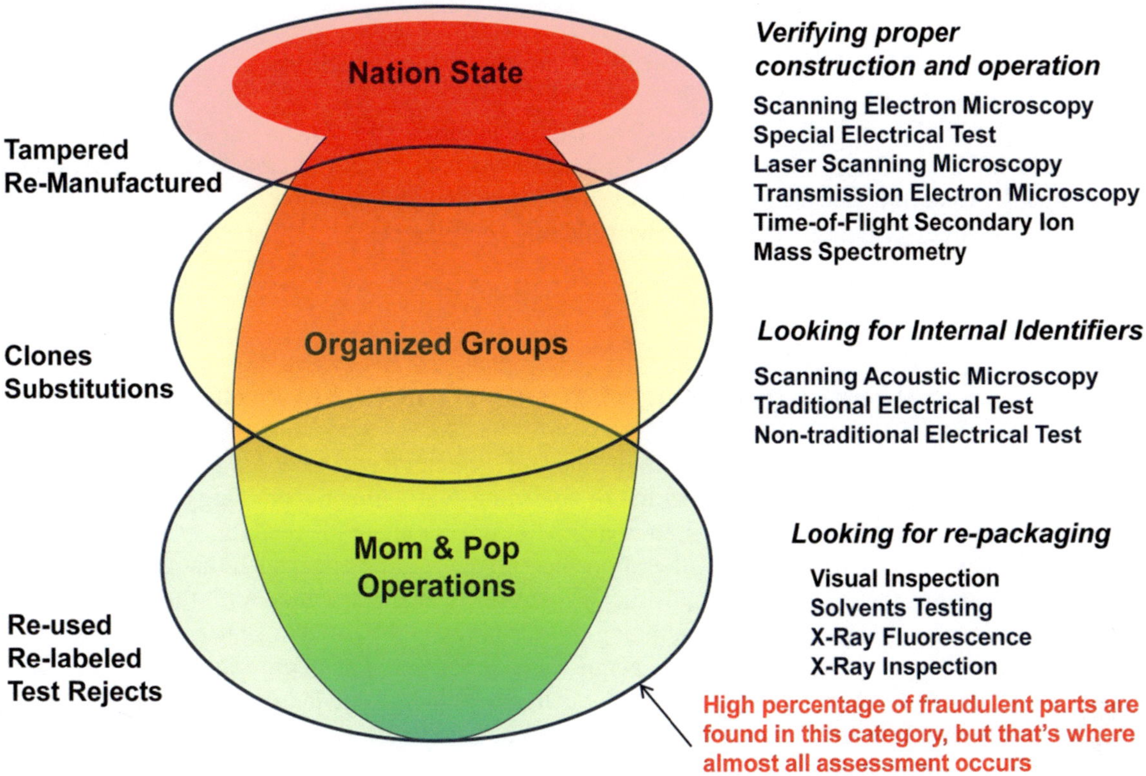

FIGURE 2-2 Technological capabilities and approaches to detecting counterfeits. SOURCE: Brett Hamilton, Chief Engineer Trusted Microelectronics, JFAC Hardware Assurance Lead, Global Deterrence and Defense Department/Flight Systems Division, Naval Surface Warfare Center, presentation to the workshop on March 16, 2016.

with weapons systems maintenance than the replacement of obsolete parts. The speaker from DARPA noted that technological solutions to ensure unchanged genuine parts and software are very possible and have the advantage of lower cost and significantly less supply-side disruption compared to bureaucratic policy solutions. In addition, DARPA noted that R&D costs are relatively inexpensive compared to added bureaucracy, and technological solutions are far easier and faster to implement. A key requirement of the DARPA SHIELD program is adoption of this taggant technique by the broader commercial industry. This is a necessary requirement to reach the cost targets and ultimate integration of this technology into the integrated circuit supply chain. Finally, one participant noted that software or hybrid software/hardware design features could help with making sure malware is not inserted in parts that are manufactured totally, or in part, in untrusted fabrication facilities; these same techniques could possibly help in the detection of clone or counterfeit parts. This "dual phenomenology" approach would make it more difficult to defeat techniques to improve trust in the supply chain.[10]

[10] Although not specifically detailed during the workshop, the following related concepts were mentioned by the participant: (1) Released firmware and software can be checked for authenticity by cryptographic methods, such as "hash" verification, which would expose any unauthorized changes to the operational code. (2) The technique described above, coupled with on-board hardware logic that would be added, would work in tandem to monitor (each other's) configuration; thus, if either the hardware, software, or firmware were modified, the combined verification check would fail. (3) Counterfeit parts would be exposed since they would not have access to, the pedigree of, or the capability to reproduce these functions.

Key Theme 3—Emerging Counterfeiting Capabilities

Several presentations (e.g., Kerry Bernstein, Brett Hamilton, and Michael Lyden) conveyed that clones and mimics are a more advanced type of counterfeit capability and an emerging concern because they are harder to detect. Accordingly, current visual inspection and common testing methods will not reveal the lack of performance expected of the authentic component.

CURRENT GOVERNMENT ACQUISITION CHALLENGES

One participant noted that the acquisition challenges for semiconductor technology can be separated into two divergent classes: (1) Class A, the acquisition of "bleeding edge" silicon technology and designs during its generation, literally at the limits of first-of-a-generation commercial availability and (2) Class B, acquisition of technology typically 3 to 4 generations behind the leading edge, such that the capital costs of obtaining such a fabricator would be a small fraction that of its original value. With these two options in consideration, this participant suggested evaluating the scenarios highlighted in Box 2-1.

A second challenge related to current acquisition processes for acquiring secure and reliable microelectronic components are relationships between government and industry program offices. One participant at the start of the workshop posed the following questions: How do we include and address rolling standards, metrics, and policies or processes, and How can any solutions be incorporated in Air Force acquisition? It was noted by some participants that the government does not necessarily know how to communicate SCRM requirements to industry, especially intelligence data on threats. A senior government leader at the workshop admitted that knowledge of SCRM requirements in government program offices is lacking and that there is a need for an integrated SCRM plan. An industry participant at the workshop strongly believed that discussions between government and industry need to occur before a contract starts and that it is critical to have engineers involved in the decision process, especially because the number of security-relevant SCRM requirements has greatly increased. (Getting these requirements into requests for proposals (RFPs) is critical.)

A participant noted that, traditionally, SCRM experts have come up through security fields, not engineering fields, and that the current thrust now is to push SCRM into systems engineering fields and acquisition fields. Another participant stated that the people who are writing the policies and acquisition RFPs also do not have these backgrounds. An industry representative stated that industry cannot do anything unless SCRM requirements are explicit in the contract—for example, common metrics for trust that are already being used by the anti-tamper community (see Figure 2-3).

One speaker on the last day of the workshop noted that, up until the 1990s, military microprocessor capabilities were superior to commercial products and that commercial products lifetimes have since been reduced dramatically. He stated that, once the trend to shorter lifetimes could begin to be observed, the government should have shifted the acquisition process to match the time-to-market shift to COTS products that were being used. A participant then posed a question, What evidence do you see of potential reforms to the acquisition system? The speaker replied that there has been some effort with respect to information systems; with respect to highly specialized defense systems, it is more difficult. A potential area to address a lot of the problems is reform of the acquisition system; although, as noted by another participant, it takes years to fight acquisition bureaucracy and to implement new practices.

BOX 2-1
Potential Technology Acquisition Approaches

Bernard Meyerson, IBM

1. *For Class A technology* the means by which trusted-by-design components can be deployed through one or more of the following approaches are the following: (1) a split foundry approach, (2) some level of trusted mask and lithography execution, or perhaps, (3) the use of autonomic monitoring of critical component behavior to provide real-time behavioral monitoring and operational assessment of the critical system element. This subset of the study does not propose to evaluate the return on investment (ROI) associated with the Department of Defense (DoD) taking ownership of a state-of-the-art fabricator.
2. *For Class B technology*, which is intended to address the needs of DoD and related agency legacy system component trust and availability, it is conceivable that fabricator ownership would meet the fiscal and technology requirements to fulfill that mission. No such outcome is assumed, but rather it is incumbent on us to perform a rigorous ROI assessment given both the dramatic devaluation of fabricator value as they age, as well as the rising costs and complexities associated with the sourcing of legacy components, the evaluation of their status as to being trusted, and the unavailability of some essential circuitry as such technology becomes obsolete.

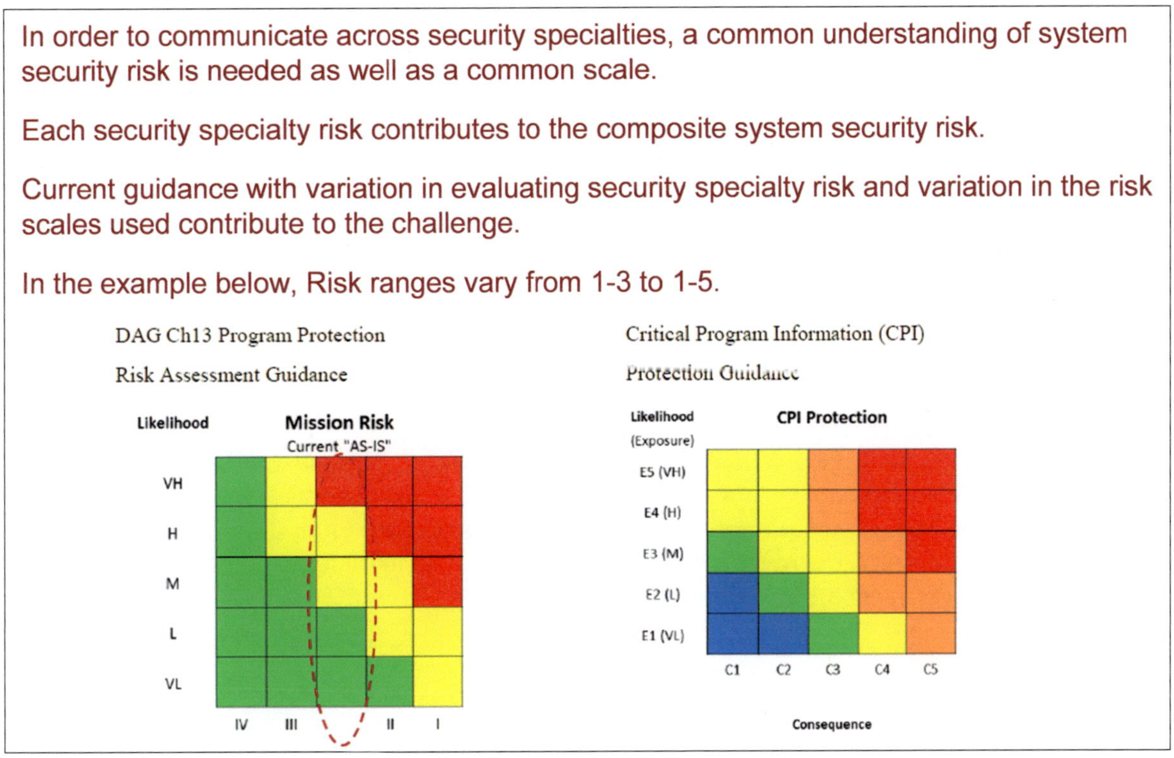

FIGURE 2-3 A proposed common government/industry approach to characterizing risk. NOTE: Mission risk cubes are widely used in the Department of Defense (DoD) systems engineering community. Typically, green signifies low risk, followed by yellow for moderate risk, followed by red for the highest risk. DAG, Defense Acquisition Guidebook. SOURCE: Holly Dunlap, Chair, Systems Security Engineering Committee, National Defense Industry Association, presentation to the workshop on March 17, 2016.

A participant noted that policy and processes are needed by the acquisition workforce to ensure that contracts, parts procurement, and methods are put into practice to detect and prevent corrupted components and vulnerabilities from entering the systems' life cycle—and one of the biggest problems is lax enforcement by government program offices of existing policies. The participant went on to specify that the acquisition process and workforce need to be more disciplined in performing program protection to assure system trustworthiness. This rigor needs to be applied to parts procurement guidelines, contracts (RFPs, statements of work, statements of objectives), design, and test. Finally, the participant stated that legacy systems and modernization programs are susceptible to bypassing recent program protection revisions to avoid extensive rework (i.e., cost) in requirements, documentation, and contracting efforts. This leads to perpetuating the fielding of vulnerable systems (which correspondingly have long operational life cycles).

Finally, another participant noted that there is far too much diversity in the rules for how DoD controls the acquisition and disposition of semiconductors and associated electronic assets. For example, in many, if not most, common systems in DoD usage today, there are system elements, such as FPGAs and graphic processing units (GPUs), that can be re-purposed after the fact. Similarly, analog circuitry has similar tuning capabilities. A critical aspect in both validating a system's correct function, as well as maintaining it over time, is the quality and trust of one's test equipment. This participant believed that it is important in any formal assessment of component acquisition that one include considerations as to the trust associated with the test equipment employed over the life of a given system and its components.

Key Theme 4—Acquisition System Implementation of DoDI 5200.44

Multiple speakers and participants (e.g., Kristen Baldwin, Brian Cohen, Harriet Goldman, and Daniel Marrujo) noted that the current acquisition system status quo is lacking in the implementation of DoDI 5200.44, which was to provide program protection for threats emanating from the supply chain and vulnerabilities in design. These speakers and participants stated that training, guidance, and security evaluation criteria need to be included in solicitations with metrics. Enforcement is needed at the program level.

Key Theme 5—Physical Limits of Current Technology

Cutting across multiple presentations (e.g., Kerry Bernstein, Carl McCants, and Bernard Meyerson) was the idea that current technology is at the end of an era as the physical limits of microelectronics have been reached (i.e., traditional scaling based Moore's Law is coming to an end). Although this is a problem for advancement for current foundries, this may be an opportunity to prepare for the next era where trust is a requirement for next-generation components.

OPTIONS FOR POSSIBLE BUSINESS MODELS WITHIN THE NATIONAL SECURITY COMPLEX

At a strategic level, OSD explained the department's planned long-term investment strategy for trusted microelectronics. The parallel components on this strategy include the following: (1) DoD identifying a commercial supplier of photomasks and building a trusted strategy to procure these; (2) transferring National Security Agency (NSA) TAPO roles and responsibilities to DMEA; (3) improving DoD microelectronics evaluation (test and validation) capabilities; and (4) developing and demonstrating alternative approaches to the trusted foundry model.[11] One participant noted that moving from a trusted hardware model to a trusted software model, which is where the department appears to be heading, is troubling—specifically, software can be made more assured, but not necessarily trusted.

The concept of split manufacturing was raised by multiple speakers as an alternative business model to the current approach by DoD (see Figures 2-4 and 2-5). Split manufacturing involves doing the initial processing steps (front end of line, or FEOL) at one foundry and finishing the fabrication at another foundry (back end of line, or BEOL). One advantage of this approach is that a higher degree of security can be obtained by doing the split earlier in the process of manufacture. The chief engineer from SMC noted that split fabrication is being reviewed by SMC as a possible alternative to loss of the current trusted foundry model.

Another possible business model for acquiring secure and reliable microelectronic components is the approach taken by the KCNSC under the Department of Energy, as summarized in Figure 2-6. Unlike the many weapon systems and technologies that DoD is responsible for, the KCNSC is responsible for acquiring and inserting ASICs in nuclear weapon systems only. In a telling remark from the speaker from KCNSC, KCNSC has since borrowed heavily from DoDI 5200.44 while implementing a formal process for SCRM. A question that was asked by one of the participants, which went unanswered, is what DoD can learn from National Nuclear Safety Administration's (NNSA's) approach that is not cost-prohibitive.

- **Demonstrate the concept of split-manufacturing of integrated circuits using a state-of-the-art offshore (untrusted) FEOL (Front End of Line) foundry and an onshore (trusted) BEOL (Back End of Line) foundry.**
  - Perform split at Metal 1 or Metal 2 – split at Metal 4 or higher used in standard split manufacturing.
  - Manage PDKs (Process Development Kits) from different foundries.
  - Fabricate chips at the 130nm, 65nm and 28nm manufacturing nodes.

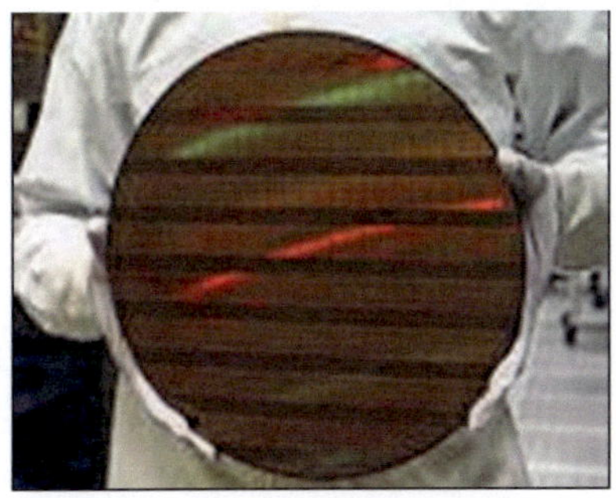

- **Develop new IC obfuscation layout strategies to protect both functional capability and performance.**
- **Anticipate and respond to evolving worldwide trends in semiconductor manufacturing.**
  - New technologies in advanced manufacturing.
  - Foundry offerings and consolidation.
  - 3D Integration (More-than-Moore).
  - Cybersecurity concerns.

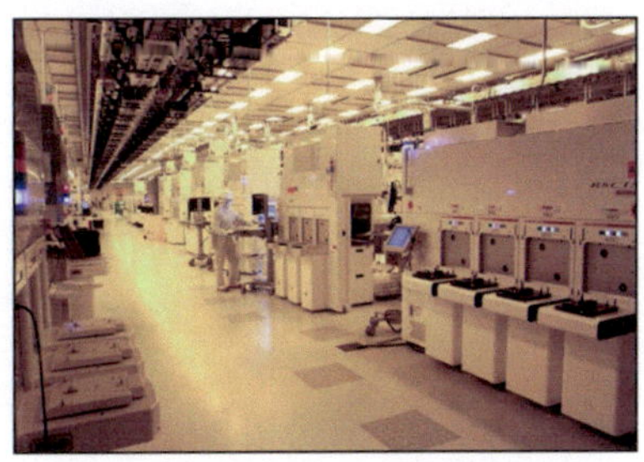

FIGURE 2-4 Trusted Integrated Circuit program approach to split manufacturing. SOURCE: Carl McCants, Program Manager, Intelligence Advanced Research Projects Activity, presentation to the workshop on March 18, 2016.

[11] Kristen Baldwin, Office of the Secretary of Defense, "Long-term Strategy for DoD Trusted Foundry Needs," presentation to the workshop on March 16, 2016.

Metrics		Phase 1A	Phase 1B	Phase 2	Phase 3
		Base (12 mo)	Option Period 1 (12 mo)	Option Period 2 (18 mo)	Option Period 3 (18 mo)
Technology Node		130 nm node	65 nm node	28 nm node	28 nm node
Circuit Complexity (# of transistors)	Digital	>10K	>100K	>1M	>10M
	Analog/Mixed Signal	>100	>1K	>1K	>10K
Split-Fabrication Yield		50%	75%	85%	95%
Speed		70%	80%	85%	90%
Power Dissipation		150%	125%	115%	110%

FIGURE 2-5 Metrics for the Trusted Integrated Circuit program. SOURCE: Carl McCants, Program Manager, Intelligence Advanced Research Projects Activity, presentation to the workshop on March 18, 2016.

In a similar fashion, this question was also asked with regard to the NSA, the founder of the original TAPO program. The NSA established the TAPO program in 2004 to provide trusted access to components used in their systems rather than recapitalize their captive integrated circuit fabrication facility they operated at that time. The NSA is no longer involved in the TAPO program, having recently turned over the management role to DMEA. The question asked was, What is NSA's current plan to ensure trusted microelectronics get used in their systems? A representative from OSD remarked that they plan to rely on their test and verification skills to validate the trust of their microelectronic components.

Other approaches for managing security and reliability risks include the following: (1) shortening the acquisition cycles, (2) aggregating microelectronics business, (3) planning for microelectronics technological change, (4) adopting commercial and industrial practices for security and reliability, (5) assessing security and reliability problems and then developing resiliency for missions and systems, and (6) developing a technological offset.[12,13] In response, a participant noted that it would be difficult to aggregate U.S. microelectronics business because of the different needs of various government agencies, although one of the main functions of TAPO (beyond ensuring access to trusted state-of-the-art parts) is aggregation of DoD's demand for trusted microelectronic components.

In the wrap-up discussions on the last day of the workshop, one participant noted that assured U.S. access to trusted microelectronic components is a pernicious problem. This participant stated that a new trusted foundry is not necessarily the answer and that there are two different, but complementary, issues—access to leading edge technology (foundries) and the issue of obsolete, counterfeit, or mimic parts. A second participant noted that the solution is not a dedicated government-run foundry and that the DoD requires many different types of electronic parts and a single foundry cannot support all these different needs (as noted above). This second participant also believed that the DARPA approach of figuring out how to build "trusted"

[12] Brian Cohen, Institute for Defense Analyses, "Obtaining Assured Electronics in a Global Commercial Marketplace," presentation to the workshop on March 18, 2016.

[13] A *technological offset* is a means of addressing a military disadvantage against either a potential or real adversary. For example, the United States developed tens of thousands of nuclear weapons during the height of the Cold War to offset the numerical advantage the Soviet and Chinese military forces enjoyed relative to Western forces.

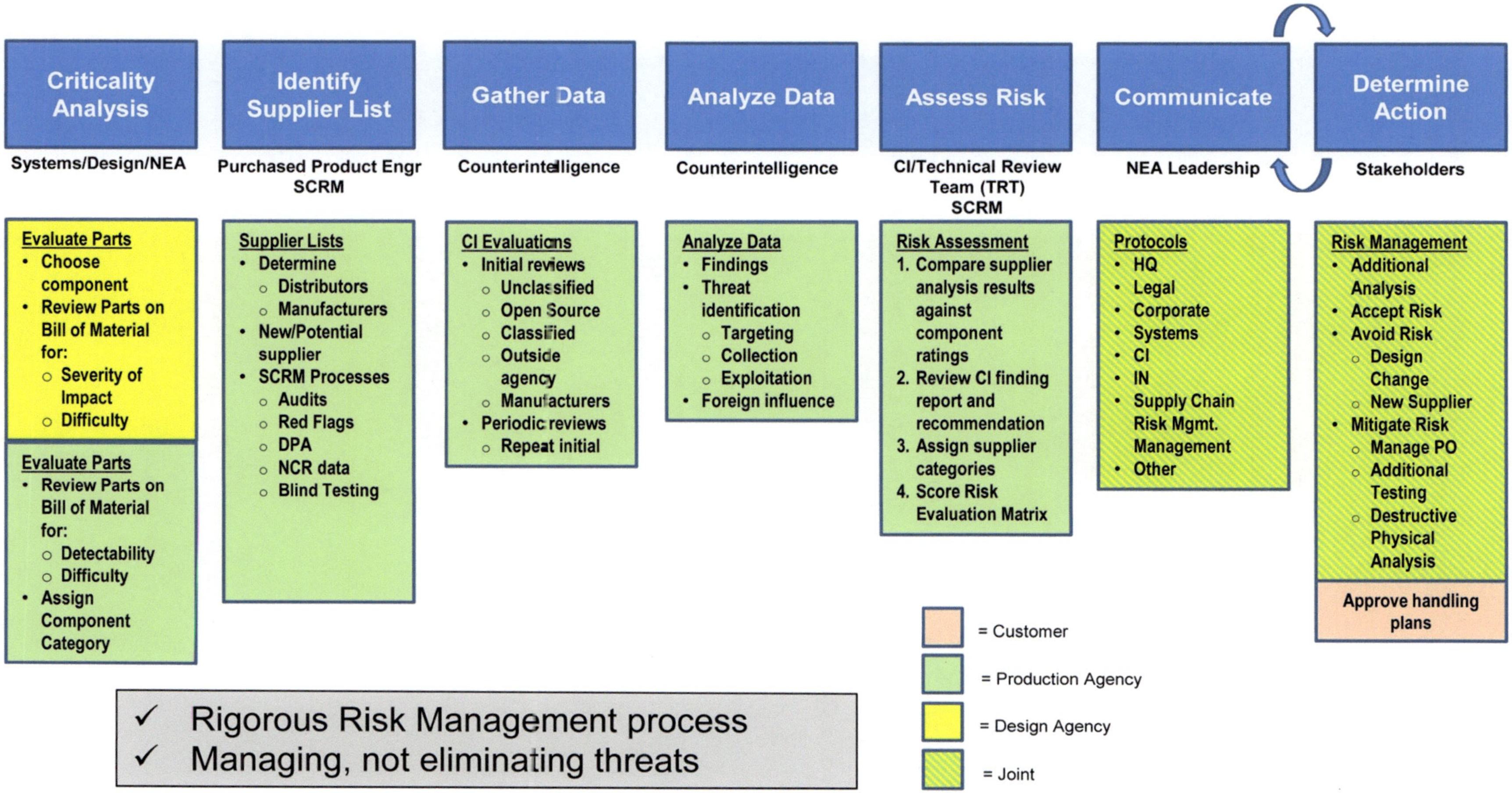

FIGURE 2-6 National Nuclear Safety Administration's (NNSA's) approach to supplier management. NOTE: SCRM, Supply Chain Risk Management. SOURCE: Kent Devenport, Technical Manager, NNSA Kansas City National Security Campus, presentation to the workshop on March 17, 2016.

integrated circuits in an untrusted supply chain is the right one. The proposed DoD strategy of seeking to extend the existing contract with GlobalFoundries in the near term to buy time, while, in parallel, making investments in both test, evaluation, and validation capabilities and in alternative approaches to the trusted foundry model (e.g., DARPA's approach) is a good one. Finally, as evidenced by the presentations from DARPA, the Intelligence Advanced Research Projects Activity, and industry, multiple new architectures and technologies exist that may provide solutions.

Key Theme 6—Trusted Foundry Model

Multiple participants (e.g., Bernard Meyerson and Michael Ettenberg) noted that the trusted foundry model is a solution to a bygone era and a new approach to assure access to trusted microelectronics may be required.

Key Theme 7—New Fabrication Methods to Replace Trusted Foundry Model

Multiple participants (e.g., Kristen Baldwin, Kerry Bernstein, Brett Hamilton, Carl McCants, and Daniel Marrujo) noted that one common vision to secure trusted components is to develop fabrication methods that ensure the microelectronics can be protected from alteration, controlled, and verified.

3

Presentation Abstracts

Listed below, in chronological order, are short abstracts or summaries of remarks provided by workshop speakers. The actual presentations were, of course, much more extensive and often covered important issues not described in the abstracts.

DAY 1—MARCH 16, 2016

Deputy Assistant Secretary of the Air Force for Science, Technology, and Engineering—David Walker (SES)

The Air Force, and the Department of Defense (DoD) in general, increasingly use application-specific integrated circuits (ASICs) to increase weapon system capability. As part of the acquisition process, the DoD must protect both the intellectual property associated with the ASIC design and the manufacturing process in order to prevent our adversaries from rapidly closing the gap of our competitive advantage, from exploiting design vulnerabilities, from sabotage, or from subversion of weapon system function. In 2004, DoD and the National Security Agency (NSA) established the Trusted Access Program Office (TAPO) to provide guaranteed access for the DoD and the Intelligence Community (IC) to trusted microelectronics technologies for their critical system needs. That same year, TAPO initiated the Trusted Foundry Program through a contract with IBM to facilitate government-wide access to trusted foundry services. Beyond the IBM contract, the Defense MicroElectronics Activity (DMEA) would accredit microelectronic suppliers as trusted suppliers. DoD formalized and consolidated its policy in 2012 and issued DoD Instruction 5200.44, which addressed supply chain risk management by requiring use of trusted suppliers for critical ASICs and implementing a program protection plan as part of the acquisition cycle.

Current State of Access to Trusted ASIC Production

Over the years, Air Force organizations and a host of programs of record used IBM and the Trusted Foundry Program to support all stages of the acquisition process from research through sustainment. TAPO renewed the Trusted Foundry Program contract in 2014. In late 2014, IBM announced its intention to sell its microelectronics business to Global Foundries, a foreign-owned entity, voiding the facility clearance license at both IBM locations used by the trusted foundry contract and breaking the trusted supply chain. As part of the Committee on Foreign Investment in the United States (CFIUS) mediation of the sale, DoD and the IC received assurances from Global Foundries that it would undertake actions to continue to provide uninterrupted trusted foundry services to the U.S. government for technologies used under the current contract until at least 2018. In addition, there are other provisions for intellectual property transfer and end-of-life notification should Global Foundries choose to shut down or discontinue a technology line.

Future Directions

There are no current alternatives to the integrated trusted foundry model offered by the Trusted Foundry Program. The Office of the Secretary of Defense (OSD) continues to work toward maintaining the Global Foundries facilities at Burlington and East Fishkill used by the Trusted Foundry Program, to enable programs to procure lifetime buys, and to negotiate with Global Foundries as a new domestic trusted supplier. Starting in fiscal year (FY) 2017, OSD will initiate a program of work to (1) establish a trusted domestic mask supplier; (2) improve DoD laboratory capability to evaluate commercial and military unique microelectronics components; and (3) develop, demonstrate, and transition technologies that enable trust by design as well as advanced evaluation capabilities.

Acting Deputy Assistant Secretary of Defense for Systems Engineering and Principal Deputy Secretary of Defense for Systems Engineering—Kristen Baldwin (SES)

For a number of years, DoD has been on a path to implement a Trusted Defense Systems Strategy. Codified in policy in 2012, "DoD acquisition programs conduct program protection planning activities throughout the life cycle to mitigate opportunities for adversaries to sabotage or subvert mission-critical system functions, system designs, and critical components of our systems. Critical components may be comprised of software, firmware, or hardware, whether specifically designed for the DoD or commercially sourced. The protection of critical components is addressed through secure engineering designs and architectures, supply chain risk management, software and hardware assurance, and anti-tamper techniques. Program protection planning gives special attention to ASICs. For ASICs that are custom-designed, custom-manufactured, or tailored for specific DoD military use, DoD requires they be procured from a trusted supplier accredited by the DMEA."[1]

"There are currently 72 DMEA-accredited suppliers, 22 of which can provide full-service trusted foundry capabilities. One of these full-service trusted foundries is Global Foundries U.S. , formerly the IBM Trusted Foundry. In addition to trust, the Trusted Foundry Program provides the U.S. government guaranteed access to leading-edge trusted microelectronics services, necessary because the low-volume DoD and Interagency needs cannot compete with commercial customers who command high-volume production requirements. The Trusted Foundry Program has served DoD and interagency needs since 2003."[2] However, this sole-source trusted foundry model carries risk, given the globalization and vertical integration of the commercial microelectronics market. Looking ahead, DoD must move to an alternative model that enables "both trust and access to needed microelectronics capability from the commercial marketplace."[3] This long-term trusted foundry strategy will improve DoD's ability to evaluate microelectronic components, protect designs from espionage or manipulation, and transition advanced technologies that permit the use of commercial sources for sensitive applications that require trust.

Defense MicroElectronics Activity—Dan Marrujo

The Trusted Foundry Program was established as a joint effort between DoD and the NSA in response to Deputy Secretary of Defense Paul Wolfowitz's Defense Trusted IC Strategy issued in 2003. The DoD component resides in OSD's Office of the Assistant Secretary of Defense for Research and Engineering. The Trusted Foundry Program is managed by DMEA. As of March 14, 2016, there are 71 Trusted Accredited

[1] Testimony of Kristen Baldwin, Assessing DoD's Assured Access to Microelectronics in Support of U.S. National Security Requirements, Hearing before the Committee on Armed Services House of Representatives, 114th Congress, 2015, H.A.S.C. No. 114-63, https://www.gpo.gov/fdsys/pkg/CHRG-114hhrg97497.

[2] Ibid.

[3] Ibid.

Suppliers offering products and services for state-of-the-art, state-of-the-practice, legacy, and obsolete microelectronics covering the entire integrated circuit supply chain.

Naval Surface Warfare Center—Brett Hamilton

Modern weapon and cyber systems are extremely sophisticated, relying on state-of-the-art electronics to achieve performance only dreamed of just a few years ago. A very high percentage of the microelectronics utilized in these systems are commercial-off-the-shelf (COTS)—many of which are designed, manufactured, packaged, and tested off-shore. Their robustness is absolutely essential to the warfighter!

Counterfeit microelectronics have been of great concern for several years now and, historically, has been widely believed to be motivated by profit. New classes of counterfeits are emerging where the motivations are not so evident. The fundamental differences between these two classes of counterfeits are highlighted below.

For Profit

Still the original part from OEM:

- Recycled used components
- Misrepresented reliability
- OEM's fab test failures sold on black market
- Unlicensed fab overproduction

Clones and Mimics

A completely different part:

- Manufactured in an unknown foundry
- Unknown process controls
- Mimics operation
- Copies based on reverse-engineering or using stolen intellectual property, potentially with altered function

This presentation will show real world examples of clones and mimics that have been examined at Naval Surface Warfare Center (NSWC) Crane Division. This will demonstrate the evolving tactics used by the counterfeiter. These tactics are very dynamic in nature, thus the tools and techniques for detection cannot be static, which presents a challenging problem for developing screening procedures. Finally the very nature of the technical assessment tools and techniques will be discussed as well as a few trends observed in the open source community.

Air Force Office of Special Investigations—Michael Lyden

The Air Force Office of Special Investigation (AFOSI) is a U.S. federal law enforcement agency that reports directly to the Office of the Secretary of the Air Force. Operating worldwide, AFOSI provides independent criminal investigative, counterintelligence, and protective service operations outside of the traditional military chain of command. AFOSI proactively identifies, investigates and neutralizes, serious criminal, terrorist, and espionage threats to personnel and resources of the U.S. Air Force and DoD, thereby protecting the national security of the United States. The desires of potential adversaries to acquire or mimic the technological advances of the U.S. Air Force have heightened the need to protect critical Air Force technologies and collateral data. The AFOSI Technology Protection Program provides focused, comprehensive counterintelligence and core mission investigative services to safeguard Air Force research and development, technologies, acquisitions, programs, critical program information, personnel, and facilities.

Defense Advanced Research Projects Agency—Kerry Bernstein

DoD's threat space for compromised sensitive electronic components is evolving quickly. Existing vulnerabilities included the counterfeiting and cloning of parts, malicious alterations, and supply chain exploits after fabrication. The recent transfer of DoD's most advanced trusted foundry to foreign ownership now introduces additional risk of intellectual property theft. For advanced lithographies, the trusted foundry era is over, and the Defense Advanced Research Projects Agency (DARPA) has been developing technologies to insure the integrity and authenticity of components used by DoD. A new methodology for asserting these tools to address specific threats faced by each component is also needed. This talk will provide an overview of tools and approaches being developed by the Microsystems Technology Office (MTO) for providing trust, which will insure that not only mission success but warfighter lives are not put at risk by compromised components.

DAY 2—MARCH 17, 2016

MITRE Corporation—Harriet Goldman

Most platform information technology systems are legacy and were designed and built prior to the nation-state cyber threats we face today. Many device manufacturers and integrators do not understand the number, or extent, of commodity- or proprietary-embedded components in their products. They are also typically unaware of the extent of hardware and software reuse, which could result in pervasive compromise across technologies and devices and cause systemic failures and cascading effects if the hardware or software is vulnerable. More importantly, many traditional cybersecurity countermeasures designed for commercial use are not adequate or even appropriate due to embedded system constraints, environmental and user considerations, and the severity of consequences. That said, the United States must respond to its eroding competitive advantage in the semiconductor space resulting in a national security risk.

Software supply chain attacks against code and application repositories through malware insertion and wide-spread code reuse and distribution are increasing (e.g., GIT hub, Mac App Store). More importantly, the Internet of Things attacks against cyberphysical and embedded systems (e.g., smart vehicles, commercial avionics, medical devices, ATMs) are becoming a reality and prominent themes at the Black Hat and RSA conferences. Attacks that disrupt the integrated circuit supply chain—whether for purposes of espionage, theft of critical data or technology, or to disrupt mission-critical operations or infrastructures—are especially nefarious. Unlike software worms or viruses, a component cannot just be wiped clean. Replacing infected hardware with a trusted component is the only option. Hardware exploits can result in adversary access and control of critical systems, cause premature or instantaneous failures in operations, or exploit cryptographic systems.

Despite recent policy and regulatory changes, heightened attention to this class of systems, and added budgeted investments, many DoD acquisition challenges remain. Some priority objective areas for focus include the following:

- *Arming program managers with better actionable threat intelligence* to better understand cyber threats to embedded microelectronics, especially hardware and the convergence of electronic warfare and cyber. Anticipatory intelligence activities to learn adversary interest and research in critical embedded system technologies can inform risk assessments and system life-cycle activities and guide investments in developing and sunsetting ineffective security and resiliency countermeasures. Similarly, reviewing whether classification, sharing policies, or practices are unduly impeding capability development and deployment should be assessed.
- *Increasing the availability of trusted countermeasures and solutions.* Guidance is lacking on the best combinations of effective protection methods (e.g., information assurance, anti-tamper, hardware assurance and software assurance, trusted suppliers, trusted foundry programs, operations security, and test and verification) for embedded systems for different missions, operating environment, and threat models. If understood, methods to develop and automate the insertion of countermeasures into hardware and firmware designs and implementations should be made a priority. In addition, approaches are needed to incentivize vendors to build these security solutions for specialized military systems (that represent a small marketplace), to create outreach programs internationally, and to

leverage innovation coming out of venture capitals, research organizations, academia, the National Laboratories, and federally funded research and development centers.

- *Creating holistic engineering and risk management practices* that minimally cover
  - Defining consistent guidance on the "How" in order to implement the "What" defined in recent directives and regulations pertinent to embedded systems;
  - Unifying often independent organizations and disciplines (e.g., mission assurance, systems engineering, security engineering, systems of systems engineering, and resiliency engineering, anti-tamper, safety critical analysis, supply chain risk management, survivability and nuclear surety) into a cohesive practice for embedded systems;
  - Shifting fundamental ideology from thinking like a defender to take the attacker's vantage point, and focusing on the adversary's goals/intent, capabilities, cyber effects, and work factor to derive security and resiliency requirements in the context of mission objectives against this threat; and
  - Righting the imbalance of guidance that exists for software to concentrate on firmware and hardware security guidance. For example, expand existing cyber frameworks and standards to cover embedded systems vulnerabilities (e.g., CVE, OVAL), threat sharing protocols (e.g., STIXTM, TAXIITM), attack patterns (e.g., ATT&CKTM), and structured languages for cyber observables (e.g., CybOXTM).
- *Automating and institutionalizing system assurance* approaches against defined metric objectives and across the systems development life cycle. Because hardware and firmware analysis is so labor intensive and expensive, automation is crucial to cost-effectively improving the quality, assurance level, and speed of the analyzing embedded components. Specifically, more best practice guidance on assurance techniques for firmware and hardware should take advantage of advancements in areas such as formal methods, side-channel analysis, fuzzy testing, encryption, trusted computing technology and trust attestation including on-chip hardware root of trust.
- *Aligning modernization efforts* with improved security. Legacy embedded system modernizations can replace insecure legacy components with newer technologies with built-in security features and lower SWaP (size, weight and power) impact. Modernizations can also support rearchitecting to minimize the attack surface and increase resilience into the future. Identifying common critical components across multiple programs and missions promotes solutions with economies of scale. Finally, opportunities to introduce innovative solutions to ride technology waves should be sought. Specifically, technology insertion roadmaps for the insertion of trusted hardware, system-on-chip components (for security), and Trojan-proof chips are needed. The ability to more frequently change the system introduces an element of surprise and uncertainty to the adversary.
- *Tracking trends, innovation, and business practices* for military advantage. Some examples to consider are the following: anticipating the "backshoring" of manufacturing, anticipating the security implications of field-programmable gate arrays programmability on security; tracking and anticipating disruptive technologies, riding technology maturity curves, and promoting legal reforms to close advantageous tax loopholes to disincentive offshoring.
- *Building capability and capacity in embedded systems security.* There is a shortage of cyber security talent in general. There is even greater capability shortfall to fill in such specialized areas as secure integrated circuit design, cyberphysical security, and reverse engineering and anti-tamper for firmware and hardware. Professional development is needed to fill this gap.

National Defense Industries Association—Holly Dunlap

Security-relevant supply chain risk management requirements are dramatically increasing. The goal to simply reduce the risk of counterfeit parts has now expanded to include component criticality analysis, malicious insertion, anonymity plans, covered defense information protection, provenance mapping, component pedigree, and trusted suppliers. A significant knowledge and awareness gap throughout the acquisition community within industry and government contributes to a barrier which stifles solutions from being integrated into systems and at times also produces overconfidence and unwarranted trust in delivered systems.

Contracts are awarded on technical merit, past performance, and cost. If security-relevant requirements are not crisply defined with metrics and measures, system security quality attributes will be traded away to system technical capability and a more affordable solution. Today, progress is being made as the presence of security-relevant requirements in contract statement of work language is increasing and maturing. However, system security and program protection have not yet made it into the contract award evaluation criteria. To encourage progress, the National Defense Industry Association (NDIA) Systems Security Engineering (SSE) Committee led a 2-year collaborative effort with the NDIA Developmental Test and Evaluation Committee, the International Council on Systems Engineering SSE Committee, the Trusted Supplier Steering Group, and MITRE to provide an industry perspective.

Air Force Space and Missile Systems Center—David Davis

Consistent with the theme of the workshop, the Space and Missile Systems Center and the broader National Security Space (NSS) systems, the current government acquisition processes for acquiring reliable and secure microelectronic components for space systems is comprehensive with numerous tenants to provide the visibility and collaboration across several fronts to ensure that an engineering, manufacturing, and test infrastructure exists, including a supply base from prime contractors through sub-tier suppliers to facilitate the development and acquisition of complex, highly reliable satellite systems, which fly in a radiation environment and we cannot perform repair on orbit.

The workshop organizing committee provided the following questions for speakers:

1. What are the current technological and government policy challenges associated with maintaining a reliable and secure source of microelectronic components?
2. What are the current government acquisition processes for acquiring reliable and secure microelectronic components?
3. What are some options for possible business models within the national security complex that would be relevant for the Air Force acquisition community with respect to secure and reliable microelectronic components?

The charts presented address the technologies and supply base critical to NSS that are necessary to engineer and produce current and future space systems that are responsive to the needed capabilities of national security. Future space systems will require leading-edge semiconductors and microelectronic devices that, in most cases, do not have a commercial market. In addition, consistent with past practices and initiatives, continued government involvement will be required to ensure a responsive industrial supply base for the products and technologies required for future space systems.

Kansas City National Security Campus—Kent Devenport

The world threat environment has changed significantly over the course of the last decade, requiring the Defense Industrial Base, including the National Laboratories and production facilities of the National Nuclear Security Administration (NNSA), to respond accordingly. Government agencies have mobilized under a variety of national-level directives to protect critical security elements against a broad spectrum of new advanced adversary threats. The U.S. government is concerned about the increased trend toward non-domestic procurement supply chain for nuclear weapon components, when coupled with the reality of increasingly sophisticated adversaries. Our defensive measures must reflect a full appreciation for the rapidly evolving, persistent, and aggressive approaches an adversary may employ that could impact our research, design, development, production, testing, storage, packaging, transportation, maintenance, surveillance, dismantlement, and disposal. The Nuclear Enterprise Assurance (NEA) program is an effort to drive activities to prevent such threats.

Kansas City National Security Campus Response

Due to current and dynamic spectrum of threats posed on the nation's Nuclear Security Enterprise (NSE), the NEA program has been established to mitigate potential consequences. NEA includes a Weapon Trust Assurance (WTA) program to ensure safe, secure and effective nuclear weapon stockpile, and a Supply Chain Risk Management (SCRM) program to ensure malicious hardware or software are prevented entry into the NSE supply chain. The underlying requirement is to design, develop, and produce all future weapons with enhanced features that are resilient to subversion attempts. This is accomplished by

1. Managing the risk of deliberate insertion of a part into the supply chain;
2. Changing the philosophy from just testing to assure functionality, to added testing to identify potential malevolent action; and
3. Working with counterintelligence to determine areas of known adversarial focus and vulnerabilities.

The Kansas City National Security Campus (KCNSC) has implemented a strong SCRM program, which includes a counterintelligence component, as well as a collaboration with other government agencies and universities to develop new technologies for trusted screenings. An awareness training program has been developed to increase the understanding of the advanced persistent threat.

IBM—Bernard Meyerson

A key question one must consider is whether or not there is a viable strategy for the U.S. government and its various departments to own, maintain, and adequately utilize a secure semiconductor foundry at a given lithographic generation. In order to make this assessment in a meaningful fashion, it is vital to first understand the trajectory existing technology is on. In approximately the year 2003, the traditional trajectory of semiconductor research, development, and manufacturing changed dramatically. Although there had been massive technological progress prior to this date, much of that progress relied upon the ongoing scaling of transistor dimensions following the admonition of Moore's Law. Roughly doubling the density of device elements on a chip every 18 months, Moore's Law provided a guide to the rate of progress in semiconductor development. However, this was enabled by a different set of rules, known as the laws of classical scaling. Classical scaling allowed one to produce a device burning exactly half the power of its predecessor, while reducing the area of the device by exactly a factor of two. This was absolutely critical, as it ensured that a chip of fixed dimension, regardless of later generation, burned precisely the same power as the prior generation, despite having twice the number of devices in its area. This relied on precisely shrinking the dimensions of all elements of the transistor. However, in 2003, a critical element of the transistor, the gate oxide, reached a dimension at which its electrical behavior became dominated by a quantum mechanical phenomena known as tunneling.

As we enter this new era in terms of what drives system performance, new opportunities present themselves to mitigate supply chain risk. We are increasingly seeing the use of field-programmable gate arrays (FPGAs) and graphic processing units as accelerative elements within systems, rather than for the ready replacement of long lead time and design intensive ASICs. It is significant that in realizing the importance of this emergent trend, Intel has acquired Altera, a leading FPGA manufacturer, and is implementing monolithic chips containing close-coupled CPUs and FPGAs having shared memory. The availability of systems on a chip with a duality of functionality makes possible real-time monitoring and validation of critical FPGA functions by an independently programmed yet close-coupled CPU. It is likely, and seen from experience, that such functionally and architecturally diverse single chips are far more robust in terms of security of function than can be achieved with a simple software- or hardware-based defense. Active methods of real-time system assurance, whether by direct monitoring as elaborated here, or via behavioral monitoring as enabled by a cognitive system exploring departures from a norm, all such options must also be explored as first or second lines of defense again malicious functionality implemented in a critical system during its manufacture.

DAY 3—MARCH 18, 2016

Institute for Defense Analyses—Brian Cohen

DoD capabilities have been repeatedly revolutionized by electronics and by the information technologies that leverage those electronics. But over time, there has been a dramatic shift in the landscape of where these technologies are developed and produced. Electronics technology and supplies increasingly come from global commercial suppliers. Innovation and manufacturing efficiency are increasingly driven by economies of scale. And these changes have resulted in both tactical and strategic risks in the supply chain. DoD has trouble obtaining specialized products at the lower volumes it needs. Low volumes of production also can compromise the yield and reliability of production. When DoD seeks out supplies, it often finds it must turn to foreign suppliers who may not provide the needed security. Even if there is security when making a buy today, the global landscape is rapidly changing, and pressures on business continue to drive industry consolidation, and there is no guarantee that important defense electronics technology and industrial capacity will be available in the United States. There are options for managing these situations in a tactical manner, but in the long term, there are some major challenges.

National Institute for Standards and Technology—Jon Boyens and Celia Paulsen

With the growing sophistication of Information and Communications Technology (ICT), along with the increased complexity of a globalized supply chain, organizations and information systems are increasingly vulnerable to supply chain risks. These risks can affect the integrity, security, resilience, safety, and quality of products and services. They may include the insertion of counterfeits into the supply chain, theft, tampering, unauthorized production, insertion of malicious code, as well as poor development practices within the supply chain.

ICT SCRM involves identifying, assessing, and mitigating risks associated with the global and distributed nature of ICT product and service supply chains. The National Institute of Standards and Technology (NIST) is responsible for developing standards, guidelines, tests, and metrics for the protection of non-national security federal information and communication infrastructure. Over the past several years, NIST has collaborated with public and private sector stakeholders to research and develop ICT SCRM tools, metrics, guidelines, and implementation strategies.

NIST's ICT SCRM program started in 2008, when it initiated the development of ICT SCRM practices for non-national security (i.e., classified) information systems, in response to Comprehensive National Cybersecurity Initiative (CNCI) #11, "Develop a Multi-Pronged Approach for Global Supply Chain Risk Management." In October 2012, NIST published NIST Interagency Report 7622, *Notional Supply Chain Risk Management Practices for Federal Information Systems,* containing a catalogue of potential ICT SCRM methods and practices centered around increasing an organization's visibility into and understanding of how the technology they acquire is developed, integrated, and deployed, thus enabling them to make risk-based acquisition decisions and develop mitigating strategies.

In 2015, NIST published NIST Special Publication (SP) 800-161, *Supply Chain Risk Management Practices for Federal Information Systems and Organizations*. This publication details a set of processes for evaluating and managing supply chain risk. These processes are integrated into the NIST SP 800-39's Risk

Management Process. Many controls in Appendix F of NIST SP 800-53 Rev. 4 can help with ICT supply chain risk mitigation. Chapter 3 of NIST SP 800-161 identifies these controls and provides supplementary guidance for their application to ICT SCRM. Additional controls assist organizations in developing more robust and complete ICT SCRM mitigation strategies. It also lists applicable threat events, provides a framework for assessing threats, and provides a template for developing ICT SCRM plans that address the entire system life cycle.

NIST is currently researching industry SCRM best practices and has published several case studies on various companies throughout different sectors of industry. In addition, NIST is working with industry, academic, and government stakeholders to identify metrics that may be useful in measuring an organization's supply chain risk. NIST is also conducting research on best practices for criticality analysis to better manage ICT supply chain risks. NIST will also begin research to demonstrate cause and effect relationships between cybersecurity and SCRM capability/maturity levels and organizational performance outcomes over time. The results will help identify which specific attributes and behaviors have disproportionate effects on cybersecurity and SCRM capability/maturity and which are more closely associated with cyber incidents.

Intelligence Advanced Research Projects Activity—Carl McCants

The semiconductor industry continues to advance rapidly with aggressive scaling and the integration of diverse analog and digital components to provide high-value microelectronic systems-on-chip. The key capabilities for fabricating the components used in these high-value systems are in commercial foundries, which now dominate the world's production of high-performance integrated circuits. It is desirable for the U.S. academic community and industrial base to have open and assured access to obtain high-performance integrated circuits and systems-on-chip, while ensuring protection of the associated intellectual property.

The goal of the Trusted Integrated Chips (TIC) program is to develop and demonstrate a new split-manufacturing process for chip fabrication, where security and intellectual property protection can be assured. The fabrication of the integrated circuit is divided into front-end-of-line (FEOL), consisting of transistor layers fabricated at an offshore foundry, and back-end-of-line (BEOL), consisting of metallization layers fabricated in trusted U.S. facilities. In this approach, the overall design intention is not disclosed to the FEOL fabricators. The development and demonstration of the TIC split-manufacturing process began at the 130 nm technology node in Phase 1A and continued at the 65 nm node in Phase 1B. For Phase 2, the TIC program performers have scaled the development of their capabilities to the 28 nm node. In Phase 3, the TIC program will explore heterogeneous split manufacturing, using a 28 nm FEOL and a 45 nm BEOL.

Appendixes

A

Terms of Reference

An ad hoc committee will be formed to facilitate an open dialogue with leading industry, academic, and government experts to:

1. Define the current technological and policy challenges with maintaining a reliable and secure source of microelectronic components;
2. Review the current state of acquisition processes within the Air Force for acquiring reliable and secure microelectronic components; and
3. Explore options for possible business models within the national security complex that would be relevant for the Air Force acquisition community.

The committee will develop the agenda for the workshop, select and invite speakers and discussants and moderate the discussions. In organizing the workshop, the committee might also consider additional topics close to and in line with those mentioned above. The meetings will use a mix of individual presentations, panels, breakout discussions, and question-and-answer sessions to develop an understanding of the relevant issues. Key stakeholders will be identified and invited to participate. One committee-authored workshop report will be prepared in accordance with institutional guidelines.

B

Committee Member Biographies

ROBERT H. LATIFF, *Chair,* retired from the U.S. Air Force as a Major General in 2006. He is a private consultant, providing advice on advanced technology matters to corporate and government clients and to universities. General Latiff is an adjunct faculty member with the John J. Reilly Center for Science, Technology, and Values at the University of Notre Dame. He is also a research professor and adjunct faculty member at George Mason University, where his interests are primarily in technologies to support the U.S. Intelligence Community. Immediately after his retirement from the Air Force, General Latiff was chief technology officer for Science Applications International Corporation's space and geospatial intelligence business. He is a member of the Air Force Studies Board of the National Academies of Sciences, Engineering, and Medicine. He has led and participated in numerous studies on such diverse topics as critical minerals, and intelligence and surveillance systems. General Latiff is an active member of the Intelligence Committee of the Armed Forces Communications and Electronics Association (AFCEA). His last active duty assignment was at the National Reconnaissance Office where he was director, Advanced Systems and Technology, and deputy director for Systems Engineering. He has also served as the Vice Commander, USAF Electronic Systems Center and Commander of the NORAD Cheyenne Mountain Operations Center. While in the U.S. Army, General Latiff served both in the infantry branch and the ordnance corps, where he commanded an Army tactical nuclear weapons unit. He received his commission from the Army ROTC program at the University of Notre Dame. He entered active service in the U.S. Army and later transferred to the U.S. Air Force. He received his Ph.D. and his M.S. in materials science and his B.S. in physics from the University of Notre Dame and is a graduate of the National Security Fellows Program at Harvard's JFK School of Government. General Latiff is a recipient of the National Intelligence Distinguished Service Medal and the Air Force Distinguished Service Medal.

MICHAEL ETTENBERG is a principal at Dolce Technologies. He was elected to the National Academy of Engineering (NAE) for contributions to the advances in optoelectronic components, including the evolution of practical and reliable semiconductor lasers. His research career included the development of some of the first commercial and reliable semiconductor lasers and the first DVD. Dr. Ettenberg was a senior vice president at Sarnoff Corporation/SRI in charge of the Solid State Division, which included integrated circuit design and foundry, microwave device and systems design and manufacture, and optoelectronics activities, including laser, LED, detector, and silicon charged-coupled device (CCD) design and manufacture. His honors and awards include the following: RCA David Sarnoff Award; Institute of Electrical and Electronics Engineers (IEEE) Third Millennium Medal; fellow of the Optical Society (OSA) and IEEE; chairman of the steering committee of the Optical Fiber Conference; past president of the IEEE Laser and Electro-Optics Society; past member of Defense Science Board; and member of the board of overseers for the New Jersey Institute of Technology. He holds a Ph.D. in materials science from New York University.

CRAIG L. KEAST is the associate head of the Advanced Technology Division at the Massachusetts Institute of Technology (MIT) Lincoln Laboratory (MIT-LL), the principal advanced electronics technology research and development division at the laboratory, since 2009. The 400-person division's focus is on the invention of new device concepts, the practical realization of those devices, and their integration into systems of importance to national security. In support of its work, the division operates and maintains a complete set of specialized microelectronic and optoelectronic fabrication facilities for both silicon and compound semiconductor devices, as well as advanced electronic and optoelectronic packaging laboratories. Program work has included split-fab fabrication activities in support of the Intelligence Advanced Research Projects Activity (IARPA) Trusted Integrated Circuit Program, the Defense Advanced Research Projects Agency (DARPA) Trusted Integrated Circuit Program, and DARPA's Integrity and Reliability of Integrated Circuits Program. From 1994 to 2013, he served as the director of the Microelectronics Laboratory (ML) where he managed operations of the laboratory's DoD-Trusted $200 million silicon-based semiconductor research and advanced prototyping fabrication facility. Staffed by ~65 scientists, engineers, and technicians working in support of more than 40 different technical programs at MIT-LL. ML activities included the fabrication of flight quality megapixel CCD imagers, photon-counting avalanche photodiode arrays, RF MEMS, Nb-based superconducting circuits, sub-0.90 nm low power FDSOI CMOS, and advanced packaging technologies. From 1996 to 2009, he was also the leader of the Advanced Silicon Technology Group, a 45-person research group carrying out work in deep-submicron, low-power, high-performance fully depleted silicon-on-insulator (FDSOI) CMOS process development, CCD/CMOS imaging, RF MEMS, Microfluidics, and 3-dimensional circuit integration technologies. From 1992 to 1994, he was a technical staff member in the Submicrometer Technology Group developing device and circuit fabrication technologies utilizing 193-nm lithography. Dr. Keast received a Ph.D. in electrical engineering and computer science from MIT.

RANDAL W. LARSON is a systems engineer with the MITRE Corporation. He has served over 40 years in engineering development and new business startups in both commercial and government sectors spanning manufacturing engineering, electrical/electronic design engineering, and systems engineering. His accumulated engineering experience includes semiconductor fabrication, electro-optic prototype development in Department of Defense (DoD) weapon systems, and design of classified, large-scale, mission-critical digital processing systems for U.S. government agencies. Additionally, he was selected as part of two technology transfer programs to launch business unit startups in enterprise-level mass storage and medical imaging systems. Mr. Larson's positions included test director, director of engineering, director of strategic planning, and general manager during these periods at Texas Instruments, Hughes Aircraft, E-Systems, and Raytheon. In 2004, Mr. Larson joined MITRE/San Antonio and was assigned to the AFLCMC/HNC "Cryptologic and Cybersecurity Systems Division (CCSD)" at Lackland Air Forece Base. During the last 12 years, roles and assignments included leading the Cryptologic Modernization Strategic Planning IPT for startup of DoD Acquisition ACAT III programs, team development of next-generation DoD Public Key Infrastructure, and Air Force research study into next-generation network security protocols and implementation of Service Oriented Architectures. In 2009 to 2010, Mr. Larson was a MITRE lead in the DoD CNCI SCRM Pilot Program for a team representing the Air Force. Follow on work for SAF/AQXA included development of SCRM roadmap and implementation for general Air Force acquisition guidance. Additionally, processes and practices were developed for implementing SCRM within the CCSD crypto acquisition programs as models for the greater Air Force. Innovative approaches included methods for evaluating DIA TAC threat reports, identifying appropriate risk mitigations, and developing a tracking database of critical components as part of establishing a TSN/SCRM office. In 2015, he assisted the director on Enterprise GPS III system (AF SMC/GPE) in establishing TSN/SCRM processes in threat/risk assessments and Program Protection planning. Mr. Larson holds a B.S.E.E. from Texas Tech.

TERRY P. LEWIS is a senior program manager and former principal systems engineer with the Raytheon Company, where his areas of expertise include command, control, communications, and information systems; digitized battlespace systems; communications and transmission security in military tactical systems; wireless network security; and network management authentication techniques for robust security architecture. In addition, Dr. Lewis has developed anti-tampering technologies to prevent or reduce the ability of potential aggressors to reverse-engineer critical U.S. communications technologies. He is a Raytheon fellow and received the Most Promising Engineer of the Year award conferred at the 2002 Black Engineer of the Year Award Conference. Dr. Lewis was a member of the Academies' Committee on Examination of the Air Force

ISR Capability Planning and Analysis Process and is a current member of the Naval Studies Board. He holds a Ph.D. in electrical engineering from the University of Southern California.

CELIA MERZBACHER is chair of the National Materials and Manufacturing Board of the Academies. Dr. Merzbacher is vice president for Innovative Partnerships at the Semiconductor Research Corporation (SRC), a nonprofit industry consortium that manages a broad portfolio of basic research on behalf of its members. She is primarily responsible for developing new initiatives and partnerships with stakeholders in government and the private sector in support of SRC's research and education mission and goals. She led the establishment of a new $10 million research effort in partnership with the National Science Foundation on Secure, Trustworthy, Assured and Resilient Semiconductors and Systems. Prior to joining SRC, Dr. Merzbacher was assistant director for technology R&D in the White House Office of Science and Technology Policy (OSTP), where she coordinated and advised on a range of issues, including nanotechnology, technology transfer, technical standards, and intellectual property. At OSTP, she oversaw the National Nanotechnology Initiative (NNI), the multiagency federal program for nanotechnology research and development. She also served as executive director of the President's Council of Advisors on Science and Technology and oversaw the council's first two statutorily mandated assessments of the NNI. Previously, Dr. Merzbacher was on the staff of the Naval Research Laboratory in Washington D.C., where as a research scientist, she developed advanced materials, including nanomaterials, for which she received six patents and authored numerous publications. Dr. Merzbacher served on the board of directors of the American National Standards Institute in relation to her role in standards development for nanotechnology. She spearheaded the establishment of the Organization for Economic Cooperation and Development Working Party on Nanotechnology and was co-lead of the U.S. delegation. She currently serves on the board of directors of Digital Solid State Propulsion, a start-up company based in Nevada. Dr. Merzbacher has served on various review committees for federal science and technology programs and advises a number of university research centers. Dr. Merzbacher holds a Ph.D. in chemistry and mineralogy from Pennsylvania State University.

BERNARD S. MEYERSON, an IBM fellow, serves as IBM's chief innovation officer, driving technical strategy and corporate initiatives within IBM's Corporate Strategy Organization. In 1980, Dr. Meyerson joined IBM Research, leading the development of high-performance silicon:germanium communications technology. He founded and led IBM's highly successful Analog and Mixed Signal business, ultimately leading IBM's global semiconductor development. In 2006, he assumed leadership of strategic alliances for the Systems and Technology Group. In 2010, he was appointed IBM Corporation's chief innovation officer, integrating his team into IBM's Corporate Strategy function, now responsible for the definition and execution of corporate-wide technical and business initiatives. Dr. Meyerson is a fellow of the American Physical Society (APS), IEEE, and a member of the NAE. His technical and business awards include the following: the Materials Research Society Medal, the Electrochemical Society Electronics Division Award, the IEEE Ernst Weber Award, the Electron Devices Society J.J. Ebers Award, the 2007 Lifetime Achievement Award from SEMI, and the 2011 Pake Prize of the APS (recognizing his combined original scientific research and subsequent business leadership). In 2014, Dr. Meyerson was honored by selection to present the Turing Lectures at the Royal Institute in London and the Universities of Cardiff, Manchester, and Edinburgh. More recently, Singapore's president honored Dr. Meyerson's service to the nation with Singapore's 2014 Public Service Medal. Most recently, in accepting a global pro-bono role, Dr. Meyerson was appointed chairman of the Meta-Council on Emerging Technologies for the World Economic Forum. In that role, he leads a diverse global team of industry, government, and university experts, the mission being the vetting and consolidation of inputs from 20 Global Agenda Councils of all major emergent technologies for presentation at the Davos meeting of the forum. He holds a Ph.D. in physics from the City University of New York.

PAUL D. NIELSEN is the director and CEO of Carnegie Mellon University's Software Engineering Institute (SEI), a federally funded research and development center sponsored by DoD. SEI develops and transitions technologies in software architecture, integration and interoperability, cybersecurity, process improvement, real time systems, and systems engineering related to software. Prior to joining SEI, Dr. Nielsen served in the

U.S. Air Force, retiring as a major general. He served primarily in research and development assignments related to space and C3I. In his final assignment, Dr. Nielsen was the commander of the Air Force Research Laboratory and the technology executive officer for the Air Force. He is a fellow of both the American Institute of Aeronautics and Astronautics (AIAA) and IEEE. He is a past president of AIAA and currently serves on the board of the Armed Forces Communications and Electronics Association. He also serves on the Defense Science Board. Dr. Nielsen received a Ph.D. in applied science from the University of California, Davis, and an M.B.A. from the University of New Mexico.

STARNES E. WALKER is the founding director of the University of Delaware Cybersecurity Initiative at the University of Delaware, with a key focus on corporate cybersecurity addressing present and emerging cyber threats and a special emphasis on the banking/financial, energy, chemical, and electrical grid industrial sectors. Previously, Dr. Walker was an executive member of the University of Hawaii System and served via an Intergovernmental Personnel Act as the chief technology officer and technical director for cyber to the U.S. Navy in a SES billet where he stood up the U.S. Fleet Cyber Command and the U.S. 10th Fleet. In this role, Dr. Walker had responsibility for all technical activities that spanned inter-governmental and international outreach of the command with a combined military and civilian workforce of 18,000 personnel. He served as a member of the Executive Steering Group to establish the Joint Technology Office-High Energy Laser Program under the auspices of the Under Secretary of Defense (Acquisition, Technology, Logistics). As a senior executive service member in helping to stand up the Defense Threat Reduction Agency, Dr. Walker was the recipient of the distinguished Department of Defense Exceptional Civilian Service Medal. He is a recipient of the R&D 100 Award and a Presidential Citation from the White House. Dr. Walker has widely published in the fields of physics, chemistry, optics, and signal processing with numerous patents issued. Dr. Walker holds a Ph.D. in physics from the University of California and an honorary degree in nuclear engineering from the University of Missouri, Rolla. Dr. Walker is a member of the Air Force Studies Board.

C

Workshop Agenda

March 16-18, 2016
The Keck Center of the National Academies
of Sciences, Engineering, and Medicine
Washington, D.C.

MARCH 16, 2016

Closed Session

0700 Breakfast (committee only)

Open Session

0800 Welcome and Introductions

Dr. Robert Latiff (Maj Gen, USAF, Ret.), Committee Chair

0815 Sponsor Expectations

Dr. David Walker (SES), Deputy Assistant Secretary of the Air Force (Science, Technology, Engineering)

0945 Break

1000 Office of the Secretary of Defense (Acquisition, Technology, Logistics)

Ms. Kristen Baldwin (SES), Acting Deputy Assistant Secretary of Defense for Systems Engineering and Principal Deputy Assistant Secretary of Defense for Systems Engineering

1100 Defense MicroElectronics Activity

Mr. Dan Marrujo, Lead MicroElectronics Reliability Engineer

1200 Working Lunch

1230 Naval Surface Warfare Center

Mr. Brett Hamilton, Chief Engineer Trusted Microelectronics, JFAC Hardware Assurance Lead, Global Deterrence and Defense Department/Flight Systems Division

1330 Air Force Office of Special Investigations

Mr. Michael Lyden, Special Agent

1430 Break

1445 Defense Advanced Research Projects Agency
Mr. Kerry Bernstein, Program Manager, Microsystems Technology Office

1545 General Discussion and Wrap Up

1600 Adjourn Open Session

Closed Session

1615 Committee Discussion

1700 Adjourn

MARCH 17, 2016

Closed Session

0700 Breakfast (committee only)

Open Session

0800 Welcome and Introductions

Dr. Robert Latiff (Maj Gen, USAF, Ret.), Committee Chair

0815 MITRE Corporation

Ms. Harriet Goldman, Director, Advanced Cyber, National Security Engineering Center, MITRE

0915 National Defense Industries Association's Systems Security Engineering Committee

Ms. Holly Dunlap, Integrated Defense Systems, Raytheon Company, Chair

1015 Break

1030 Air Force Space Command/Space and Missile Systems Center

Mr. David Davis, SMC Chief Systems Engineer

1130 National Nuclear Security Administration

Mr. Ken Devenport, Technical Manager, Kansas City National Security Campus, Department of Energy

1230 Working Lunch

1300 IBM

Dr. Bernard Meyerson, Chief Innovation Officer

1400 The Aerospace Corporation

Dr. Allyson Yarbrough, Principal Engineer, Electronics and Sensors Division

1500 Break

Closed Session

1515 Committee Discussion

1700 Adjourn

MARCH 18, 2016

Closed Session

0700 Breakfast (committee only)

Open Session

0800 Welcome and Introductions

Dr. Robert Latiff (Maj Gen, USAF, Ret.), Committee Chair

0815 Institute for Defense Analyses

Dr. Brian Cohen, Research Staff Member, Information Technology and Systems Division

0915 National Institute for Standards and Technology

Mr. Jon Boyens, Project Lead, SCRM for Information and Communications Technology
Ms. Celia Paulsen, Technical Lead, SCRM for Information and Communications Technology

1015 Break

1030 Intelligence Advanced Research Projects Activity

Dr. Carl McCants, Program Manager

1130 Feedback from Sponsor on Next Steps

Dr. David Walker (SES), Deputy Assistant Secretary of the Air Force (Science, Technology, Engineering)

Closed Session

1200 Committee Discussions on Key Themes *with Lunch Available*

1500 Adjourn

D

Workshop Attendees

COMMITTEE MEMBERS

Dr. Robert H. Latiff (Maj Gen, USAF, Ret.), R. Latiff Associates, *Chair*
Dr. Michael Ettenberg, Dolce Technologies
Dr. Craig L. Keast, MIT Lincoln Laboratory
Mr. Randal W. Larson, MITRE Corporation
Dr. Terry P. Lewis, Raytheon Company
Dr. Celia Merzbacher, Semiconductor Research Corporation
Dr. Bernard S. Meyerson, IBM
Dr. Paul D. Nielsen (Maj Gen, USAF, Ret.), Software Engineering Institute
Dr. Starnes E. Walker, University of Delaware

ACADEMIES STAFF

Dr. Joan Fuller, Director, Air Force Studies Board
Mr. Carter W. Ford, Program Officer, Air Force Studies Board
Mr. Steven Darbes, Research Assistant, Air Force Studies Board
Ms. Marguerite E. Schneider, Administrative Coordinator, Air Force Studies Board

SPEAKERS

Dr. David Walker (SES)
Deputy Assistant Secretary of the Air Force (Science, Technology, Engineering)

Ms. Kristen Baldwin (SES)
Acting Deputy Assistant Secretary of Defense for Systems Engineering and Principal Deputy Assistant Secretary of Defense for Systems Engineering

Mr. Kerry Bernstein
Program Manager, Microsystems Technology Office, Defense Advanced Research Projects Agency

Mr. Jon Boyens
Project Lead, SCRM for Information and Communications Technology, National Institute for Standards and Technology

Dr. Brian Cohen
Research Staff Member, Information Technology and Systems Division, Institute for Defense Analyses

Mr. David Davis
Chief Systems Engineer, Space and Missile Systems Center, Air Force Space Command

Mr. Ken Devenport
Technical Manager, Kansas City National Security Campus, Department of Energy, National Nuclear Security Administration

Ms. Holly Dunlap
Integrated Defense Systems, Raytheon Company, and Chair, NDIA Systems Security Engineering Committee

Ms. Harriet Goldman
Director, Advanced Cyber, National Security Engineering Center, MITRE

Mr. Brett Hamilton
Chief Engineer Trusted Microelectronics, JFAC Hardware Assurance Lead, Global Deterrence and Defense Dept/Flight Systems Division, Naval Surface Warfare Center

Mr. Michael Lyden
Lead Analyst, Technology Protection Team, Air Force Office of Special Investigations

Mr. Dan Marrujo
Lead MicroElectronics Reliability Engineer, Defense MicroElectronics Activity

Dr. Bernard Meyerson
Chief Innovation Officer, IBM

Dr. Carl McCants
Program Manager, Intelligence Advance Research Projects Activity

Ms. Celia Paulsen
Technical Lead, SCRM for Information and Communications Technology, National Institute for Standards and Technology

Dr. Allyson D. Yarbrough
Principal Engineer, Electronics and Sensors Division, The Aerospace Corporation

GUESTS

Mr. Matthew Casto
Senior Electronics Engineer, Air Force Research Laboratory

Mr. Richard-Duane Chambers
Briefing Coordinator, Booz Allen Hamilton

Mr. Patrick Cheetham
Research Associate, Potomac Institute for Policy Studies

Mr. Dean Collins
Managing Member, DRC Consulting LLC

Mr. Barry Davilli
Princ Systems Engineer, SCRM, Raytheon Corporation

Dr. Michael Fritze
Senior Fellow, Potomac Institute for Policy Studies

Mr. Jimmy Goodrich
Vice President, Global Policy, Semiconductor Industry Association

Mr. Joseph Gordon
Air Force S&T Management Division Chief, Office of the Deputy Assistant Secretary of the Air Force (Science, Technology, Engineering)

Ms. Jennifer Lato
Research Associate, The Potomac Institute for Policy Studies

Mr. Gabriel Mounce
Senior Electronics Engineer, Space Vehicles Directorate (Space Electronics Program)
Air Force Research Laboratory

Mr. Anthony Newton
C4I and Cyber PEM, Office of the Deputy Assistant Secretary of the Air Force (Science, Technology, Engineering)

Mr. P. Len Orlando III
Engineer, Air Force Research Laboratory

Mr. Raymond Shanahan
Deputy Director, Anti-Tamper/Hardware Assurance, Office of the Deputy Assistant Secretary of Defense (Systems Engineering)

Mr. Dustin Todd
Director, Government Affairs, Semiconductor Industry Association

Mr. Joseph E. Van Nostrand
Senior Electronics Engineer, Information Directorate, Air Force Research Laboratory

Mr. Glen D. Via
Principle Electronics Engineer, Air Force Research Laboratory

Mr. James A. Will II
Principal Engineer, Department of Energy, NNSA's National Security Campus

E

Potential Terms of Reference for Follow-on Study

During the course of the 3-day workshop the Air Force sponsor and other participants asked the question, Is there value in conducting a future National Academies of Sciences, Engineering, and Medicine study to pursue the topics raised throughout the workshop in greater detail? Box E-1 provides notional terms of reference authored by the workshop committee for future consideration.

BOX E-1
Terms of Reference

The National Academies of Sciences, Engineering, and Medicine will appoint a study committee to conduct a consensus study in accordance with Academies procedures. The Academies will then:

1. Review and describe current Air Force acquisition policies and requirements for secure and reliable microelectronic components. Compare these with approaches used by other Services, the Intelligence Community, and industry.
2. Identify and describe Air Force capabilities requiring secure and reliable microelectronic components.
3. Identify and describe the current and forecasted (on a 5-year horizon) range of threats to the supply chain.
4. Identify and describe acceptable levels of trust required for those Air Force capabilities identified as requiring secure and reliable microelectronic components.
5. Recommend ways to resource and institutionalize future Air Force acquisition of secure and reliable microelectronic components.

A substantive unclassified report, which addresses the terms of reference and may include a classified appendix, will be produced no later than 12 months after receipt of funding.

F

Projected Advancements of Existing Technology

During the planning phase of the workshop, the organizing committee considered the key question of whether or not there is a viable strategy for the U.S. government and its various departments to own, maintain, and adequately utilize a secure semiconductor foundry at a given lithographic generation. Several members of the organizing committee believed that in order to make this assessment in a meaningful fashion, it would likely require understanding the trajectory existing technology is on as well as understanding the economic landscape that resulted in the collapse of the U.S. electronics manufacturing sector. For the military community, Bernard Meyerson offered to summarize his views and share with the organizing committee and workshop participants. He stated that in approximately 2003, the traditional trajectory of semiconductor research, development, and manufacturing changed dramatically. Although there had been massive technological progress prior to this date, much of that progress relied on the ongoing scaling of transistor dimensions following the trend known as Moore's Law. Predicting that the number of transistors on a chip will roughly double every 18 months, Moore's Law provided a guide to the rate of progress in semiconductor development. However, this was enabled by a different set of rules, known as the laws of classical scaling (see Figure F-1).

Classical scaling allowed one to produce a device burning exactly half the power of its predecessor, while reducing the area of the device by exactly a factor of two. This was absolutely critical, as it in ensured that a chip of fixed dimension, regardless of later generation, burned precisely the same power as the prior generation, despite having twice the number of devices in its area. This relied on precisely shrinking the dimensions of all elements of the transistor. However, in 2003, a critical element of the transistor, the gate oxide, reached a dimension at which its electrical behavior became dominated by a quantum mechanical phenomena known as tunneling.

Effectively, the previously insulating gate oxide layer had been rendered a useless conductor. This was the beginning of the end as to the performance benefits derived solely by scaling of the following generations of silicon technology. The impact is seen in Figures F-2 and F-3.

By virtue of the cost of the technological innovations required to mitigate such phenomena, this triggered the economic collapse of any subcritical scale commercial effort in silicon technology, resulting in a handful of leading-edge foundries surviving this transformation of the industry. With an inability to achieve material performance gains by the simple scaling of an existing silicon generation, the industry needed to resort to extremely costly innovations—in the materials used, the processing employed, device geometry, substrate materials, and a host of other elements—in producing the following generations. As represented in Figure F-3, this led to the rapid escalation in the cost of developing each subsequent generation of technology, similarly resulting in a rapid falloff in the number of vendors choosing to continue to pursue this strategy. The consolidation of this industry continues even today, and the complexity of the issues raised here has only become ever more problematic.

In recent technology generations, the benefits of next-generation technology for the actual performance of a single processing thread and/or core in a microprocessor have become essentially nil, as seen in Figure F-4. Compensating for this, more cores and other assets on a die have been implemented to improve performance at a system level (functionality/$), but in terms of raw performance from devices themselves, that benefit has gone asymptotically to zero.

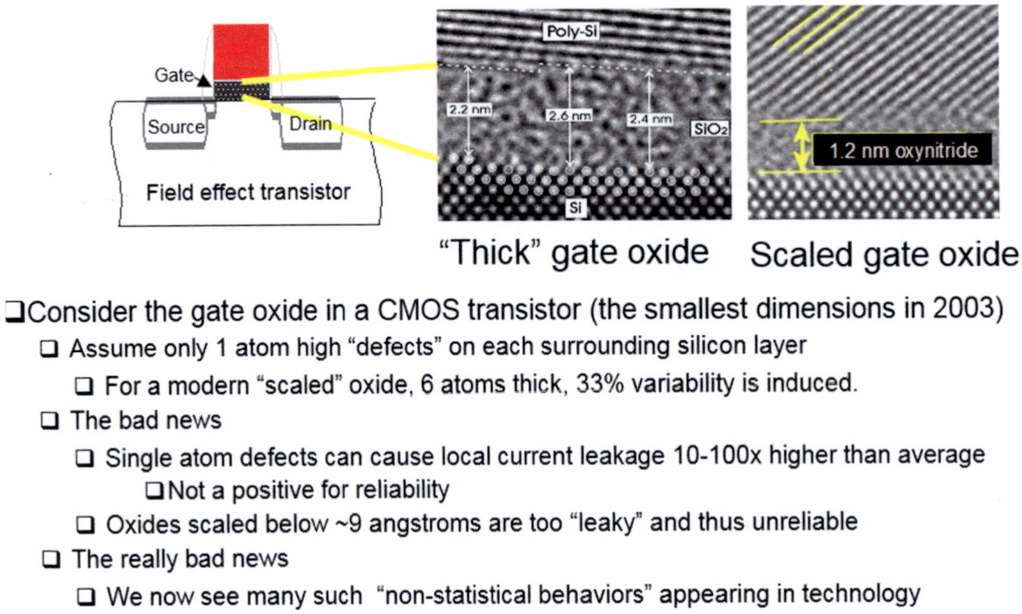

FIGURE F-1 The 9 versus 10 angstrom "debate" from 2003. SOURCE: Bernard S. Meyerson, "Driving System Performance—A New Paradigm (For most technologists)," presentation to the 2004 Microprocessor Forum, San Jose, Calif.

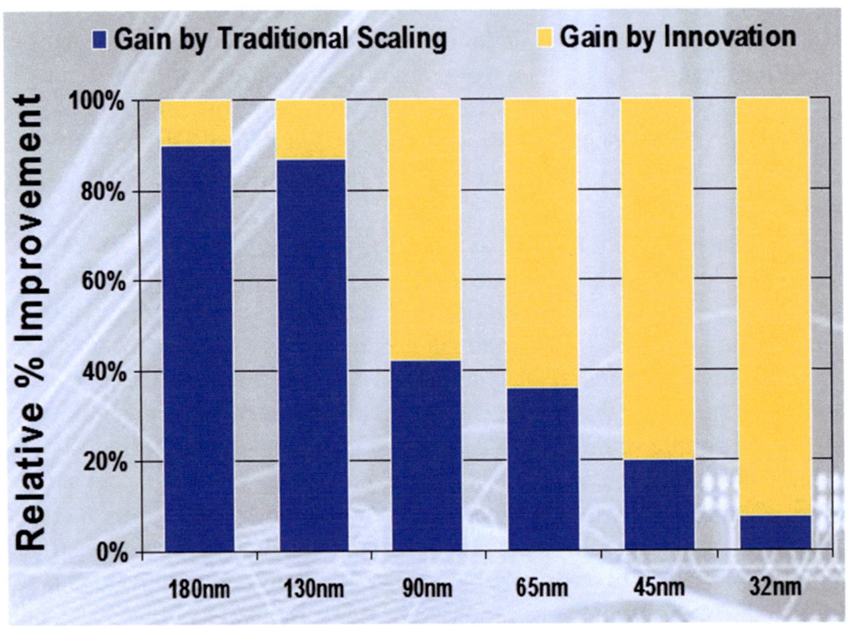

FIGURE F-2 Green computing. SOURCE: Bernard S. Meyerson, International Symposium on Semiconductor Manufacturing (ISSM) 2008 Keynote.

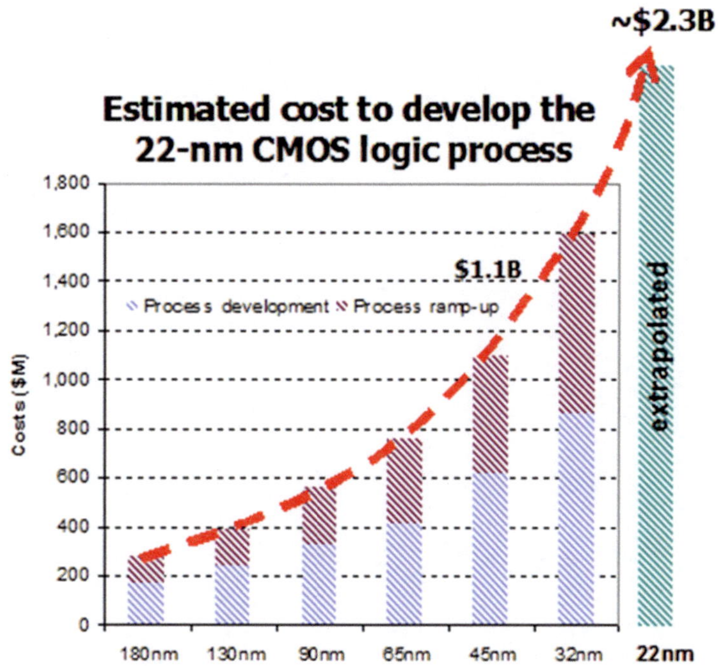

FIGURE F-3 Rapid escalation in the cost of developing subsequent generations of technology. SOURCE: Bernard S. Meyerson, "Echoes of DACs Past: From Prediction to Realization, and Watts Next?" Design Automation Conference, June 2010, Anaheim Calif. From IBS Global System IC Service Management Report, April 2006.

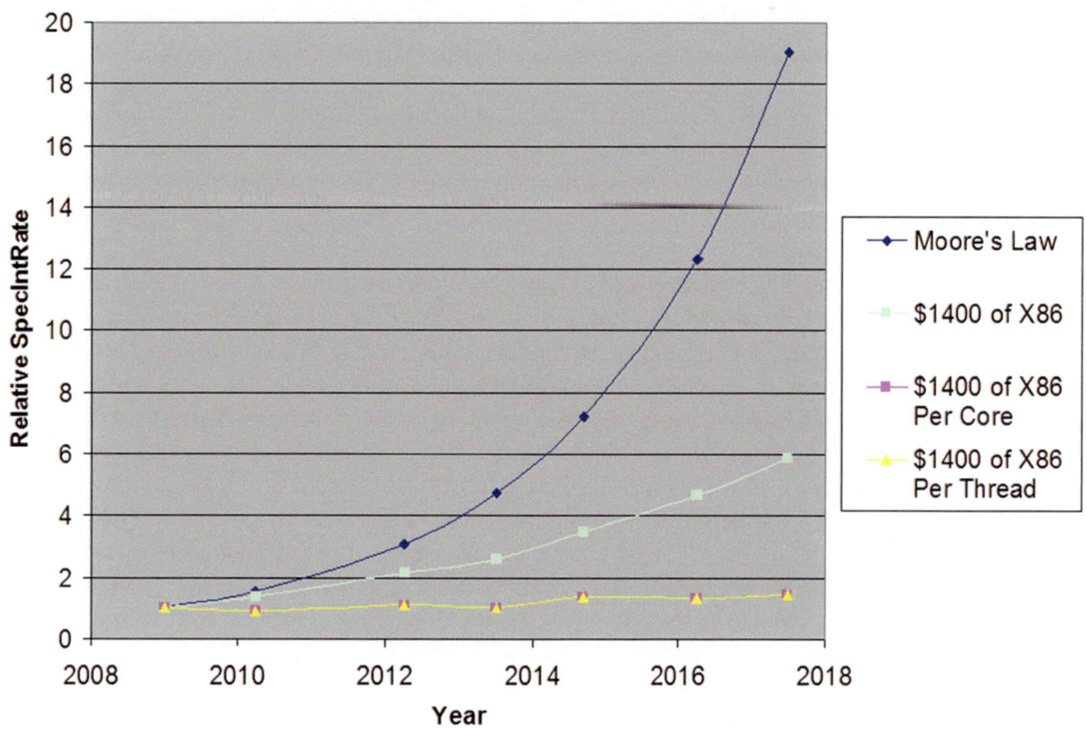

FIGURE F-4 Performance (SpecIntRate) over time; IBM estimates, 2016. SOURCE. Bernard S. Meyerson.

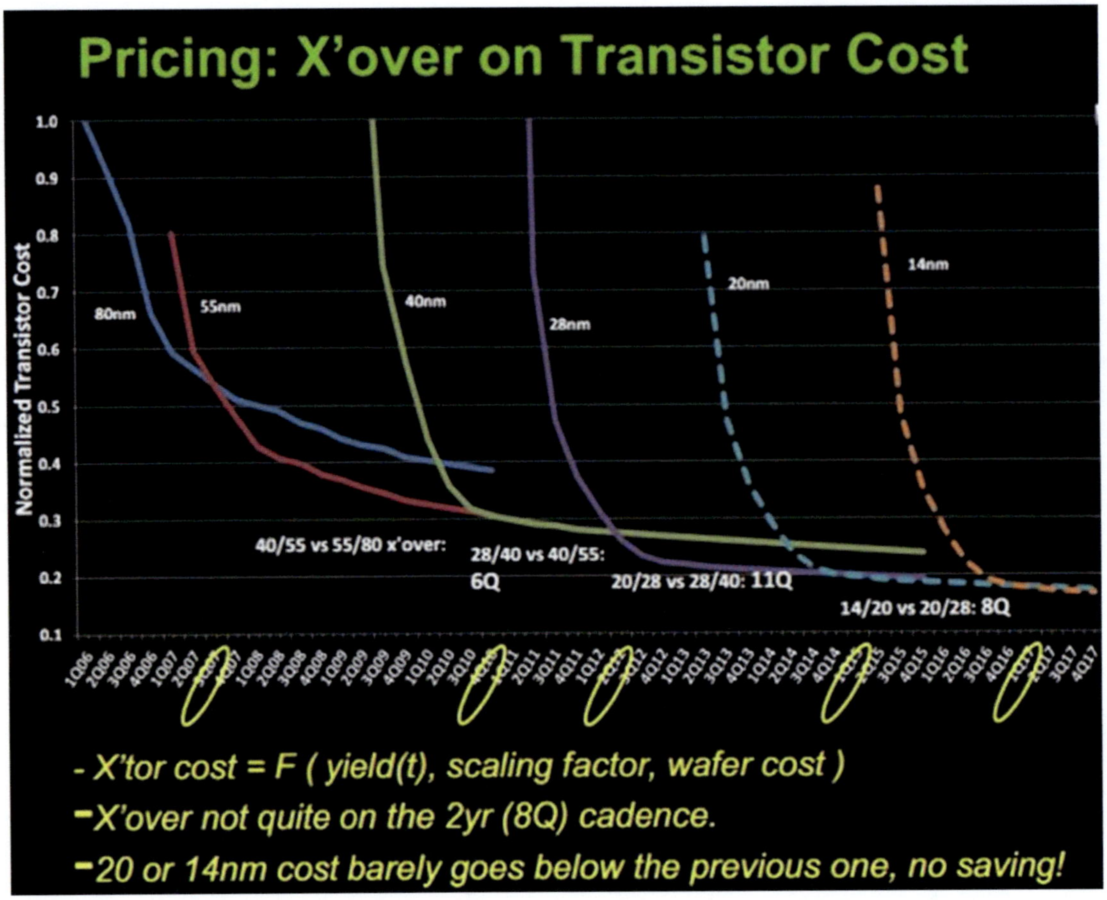

FIGURE F-5 Benefits for power and performance and cost per function. SOURCE: From Jen-Hsun Huang, NVIDIA, presentation to the International Trade Partner Conference, November 2011, available at http://www.extremetech.com/computing/123529-nvidia-deeply-unhappy-with-tsmc-claims-22nm-essentially-worthless. Courtesy of NVIDIA.

There are still clear benefits for power and performance and cost per function as represented by the increase in performance for a fixed dollar invested in X86 systems (see Figure F-5). However, to emphasize the implications of this long term, NVIDIA created the data in Figure F-5 from preliminary pricing and yield curves for past and future generations of silicon technology, as provided by NVIDIA, a fabless producer of graphics chips, and highlighted that over time, the drive to move to new generations goes away as future costs of manufacturing cause the cost per transistor for new generations to remain equal to or above that of prior generations (28 nm → 20 nm → 14 nm). This eliminates the key economic driver of movement to the next generation of technology, with implications as to the rate, pace, and economic success of this industry.

What is of concern is that for first time, the price per transistor of a following generation fails to fall below that of the prior generation, raising the issue of why one would move to the new generation given no improvement in the economics and little improvement in overall performance. It is not that there are no gains to be had, it is simply that they are asymptotically approaching zero.

With benefits of further scaling ever less with time, a more definitive statement can still be made. Silicon itself becomes the limiting material in the near term, at which time further scaling of any sort ceases to be viable. This is due to silicon itself being rendered un-usable when it approaches quantum mechanical limits at somewhere in the range of 4 to 7 nm. In this regime, even one or two atomic width deviations in device dimensions will impart so much variability to devices as to make a circuit difficult, if not impossible, to produce in a controllable fashion. Compounding this problem, the metallization utilized for connectivity in such devices does not scale as well as silicon itself, also becoming a final gating factor. Therefore, it can be argued that it is unwise for the government to even consider an option under which it might acquire and operate a secure foundry at the very leading edge of technology. This is unwise from a technical, as well as financial, perspective. Such a leading-edge foundry could cost in excess of $10 billion to develop, and to be viable, it would have to be operated at virtually 100 percent utilization at all times. There is no volume at the leading edge of technology within the Department of Defense (DoD) or other agencies that would remotely fill even a small fraction of such a capacity, and the maintenance of such a facility at low-volume production is virtually impossible, should one want to ensure quality and process stability.

By contrast, back level foundries rapidly become legacy assets, so the government could expect to reasonably acquire such a relatively current (n-2,4) foundry for a dramatically discounted capital expense. However attractive this may seem, a full return on investments and return on assets analysis would be required to validate such an approach as sustainable. Although the initial capital expense would be minimized, the challenge remains in the overall operation and associated investment costs for such an endeavor. This comprehends everything from the creation of physical design kits, to the instantiation of a design flow, to the creation of what would effectively be a foundry support organization. This is not to presuppose success or failure of this analysis, but it is critical to comprehend what all associated costs will be upon the acquisition of the foundry in order to make such a judgment. Whereas previously one might dismiss this notion out of hand due to the rapid movement of silicon technology into the future, the asymptotic approach to silicon's "end of life" in terms of further scaling greatly mitigates the rate at which such an asset would become such a legacy so as to burden its users.[1] Having made these assertions to the point where further detailed analysis is required, it is also worth turning our attention to other means by which a secure supply line may be maintained without actually acquiring the large and ongoing challenges associated with semiconductor manufacturing.

As we enter this new era in terms of what drives system performance, new opportunities present themselves to mitigate supply chain risk. We are increasingly seeing the use of field-programmable gate arrays (FPGAs) and graphic processing units (GPUs) as accelerative elements within systems, rather than for the ready replacement of long lead time and design intensive application-specific integrated circuits (ASICs). It is significant that in realizing the importance of this emergent trend, Intel has acquired Altera, a leading FPGA manufacturer, and is implementing monolithic chips containing close-coupled CPUs and FPGAs having shared memory. Further, Xilinx has produced the next generation of system on module (SOM), which couples programmable logic with embedded ARM hard CPU cores. The availability of systems on a chip with a duality of functionality makes possible real-time monitoring and validation of critical FPGA functions by an independently programmed yet closely coupled CPU. It is likely, and seen from experience, that such functionally and architecturally diverse single chips can be more robust in terms of security of function than can be achieved with a simple software- or hardware-based defense. Active methods of real-time system assurance, whether by direct monitoring as elaborated here, or via behavioral monitoring as enabled by a cognitive system exploring departures from a norm, are options to be explored as first or second lines of defense against malicious functionality implemented in a critical system during its manufacture. Unfortunately, functionality running in so-called bare metal configurations runs at clock-rate speeds, whereas software does not. The implications are that real-time checking of hardware may be difficult to perform by software; however, hardware can be used to verify and validate correctness of software.

---

[1] It may be that as silicon technology hits the end-of-life wall, there may ultimately be *more* suppliers who will be able to reach this capability limit in a cost effective way. Silicon fabrication will truly be a "commodity process" at this point, and not a differentiator on product performance.